T0271970

Stochastic Analysis for Gaussian Random Processes and Fields

With Applications

MONOGRAPHS ON STATISTICS AND APPLIED PROBABILITY

General Editors

F. Bunea, V. Isham, N. Keiding, T. Louis, R. L. Smith, and H. Tong

Monographs on Statistics and Applied Probability 145

Stochastic Analysis for Gaussian Random Processes and Fields

With Applications

Vidyadhar S. Mandrekar

Michigan State University

East Lansing, USA

Leszek Gawarecki

Kettering University

Flint, Michigan, USA

CRC Press
Taylor & Francis Group
Boca Raton   London   New York

CRC Press is an imprint of the
Taylor & Francis Group, an **informa** business

A CHAPMAN & HALL BOOK

CRC Press
Taylor & Francis Group
6000 Broken Sound Parkway NW, Suite 300
Boca Raton, FL 33487-2742

© 2016 by Taylor & Francis Group, LLC
CRC Press is an imprint of Taylor & Francis Group, an Informa business

No claim to original U.S. Government works

International Standard Book Number-13: 978-1-4987-0781-7 (Hardback)

Library of Congress Cataloging-in-Publication Data

Mandrekar, V. (Vidyadhar), 1939-
    Stochastic analysis for gaussian random processes and fields : with applications / Vidyadhar S. Mandrekar and Leszek Gawarecki.
        pages cm. -- (Monographs on statistics and applied probability ; 145)
    "A CRC title."
    Includes bibliographical references and index.
    ISBN 978-1-4987-0781-7 (alk. paper)
    1. Gaussian processes. 2. Stochastic processes. I. Gawarecki, Leszek. II. Title.

QA274.4.M37 2015
519.2'4--dc23   2015007941

Visit the Taylor & Francis Web site at
http://www.taylorandfrancis.com

and the CRC Press Web site at
http://www.crcpress.com

*We dedicate this book to the memory of
Professor R.R. Bahadur
and to the mentor of V. Mandrekar
Professor S.D. Chatterji*

Contents

Preface

The book studies the Gaussian random fields using the structure of the Hilbert space of functions on the parameter space associated with the covariance, the so-called Reproducing Kernel Hilbert Space (RKHS). The RKHS associated with the covariance kernel was first used by Aronszajn [3]. It was used in the context of equivalence and singularity of Gaussian Random Fields by Parzen [103] and Kallianpur and Oodaira [55]. Their idea was exploited to provide a simplified presentation of the problem in [15]. The beauty of the technique is in using methods of Hilbert space to study deeper analytic properties connecting probabilistic notions. A good example of this approach is the work of Skorokhod [118], which studies differentiation on the space of random variables. Further application in the analysis of differentiation was achieved by Malliavin, who introduced this concept independently in his fundamental work [76] using what is now called the Malliavin derivative. Using the ideas of Nualart [94], [95], we show that the two concepts coincide. For deeper analysis, one can see the publications of Bismut [6]. The work of Kallianpur and Mandrekar [81] on Lévy Markov property connects it to the analytic structure of RKHS of the Gaussian random field. This explains the free field Markov process of Nelson [90] and local operators of Kusuoka [69]. Our purpose in this book is to start with the study of RKHS of Aronszajn and associating a Gaussian subspace with the RKHS, as in Kakutani [51], to explain the concepts introduced above. In addition, we present the applications of these ideas to current research in the areas of finance, spatial statistics, and the filtering and analytic problem related to fractional Brownian motion. The earlier development of some chapters originated in [79].

The book starts with the presentation of preliminary results on covariance and associated RKHS needed in Chapter 1. We then introduce the Gaussian process as a map π between RKHS $K(C)$ and a subspace of square integrable functions on a probability space, $L^2(\Omega, \mathcal{F}, P)$. This gives a natural definition of the Wiener integral as in [80]. It is then easy to obtain the integral representation of certain Gaussian processes. We end Chapter 2 by presenting the definition of multiple Wiener integrals for a general Gaussian process as an extension of the map π to tensor products $K(C)^{\otimes n}$ of RKHSs. As a consequence we give a representation of elements of $L^2(\Omega, \mathcal{F}, P)$ in terms of multiple Wiener integrals (chaos expansion). This work is based on the approach in [84].

The chaos expansion is used in Chapter 3 to define the Skorokhod integral, which generalizes the Itô integral. We then define the Ogawa integral following [31], which generalizes the Stratonovich integral. In addition, to relate the ideas of Malliavin and Skorokhod we present the concept of Skorokhod differentiation, and in Chapter 4 we show that, in the case of Brownian motion, it coincides with the Malliavin derivative. Therefore, the Skorokhod integral is a dual operator of Skorokhod differentiation and thus it is the divergence operator of Malliavin. Our presentation follows the ideas in [94] and [95]. We end Chapter 4 by showing the role of stochastic differentiation in finance by discussing the concept of stochastic duration [57]. As an application of our approach, we derive the recent results of [64] on the Itô formula for Gaussian processes with a special structure of covariance with application to fractional Brownian motion.

In Chapter 5, we study Gaussian processes indexed by real numbers and obtain a Kallianpur–Striebel Bayes formula for the filtering problem with the measurement error being a general Gaussian process following [77]. In a particular case, which includes fractional Brownian motion, we derive the analogue of the Zakai equation. One can then solve the filtering problem, including Kalman filtering, using fractional Brownian motion noise. The techniques used are based on the work in [33] and [35].

We consider the problem of equivalence and singularity of Gaussian random fields in Chapter 6. As in [14], we obtain general conditions for equivalence and singularity in terms of the relation between RKHSs associated with the covariances of two fields. In a special case of stationary random fields, we use these results to obtain spectral conditions used by Stein [120] in the problem of interpolation of stationary Gaussian random processes and analogue results in [46] for stationary random fields. The approach is as in [19].

At the end of the chapter, we give a generalization of the Girsanov theorem for Gaussian random fields and derive other results in this direction as consequences following [32].

In Chapters 7 and 8, we study the Markov property of Gaussian random fields indexed by measures and generalized Gaussian random fields indexed by Schwartz space. This part is motivated by the work of Nelson [90], Dynkin [21], and Röckner [108]. However, our approach is through general conditions on RKHS of a Gaussian random field using the ideas of Molchan [88] and Kallianpur and Mandrekar [81]. To derive the results in [21] and [108], we use the techniques from [84] and this is accomplished as in [124]. In order to present the concepts involved, we needed to present some results on Dirichlet forms and associated Markov processes from Fukushima [30]. For the convenience of the reader, these are presented in Appendix 7.4 to Chapter 7. The Gaussian random field is associated with Green's function of the Markov process and is indexed by measures. The terminology is also explained.

In the final Chapter 8, we first derive from our general results the work of

Pitt [104] and Künch [66] on Markov property of Gaussian fields indexed by $\mathbb{R}^d$. In addition, we relate the general work on measure–indexed random fields with general conditions associated to the Dirichlet form generating the Markov process. This is a recent work of Albeverio and Mandrekar [2].

We regret that we cannot present here the interesting results of Okabe [101], Kotani [62], and Pitt [105] on Markov property of stationary Gaussian random fields with analytic conditions on spectral density. This involves additional technical structures from complex analysis, like, for example, ultra distributions. However, we refer the interested reader to the unpublished technical report by Mandrekar and Soltani [83].

Vidyadhar S. Mandrekar
Leszek Gawarecki

Acknowledgments

We want to thank Professors S. Albeverio, S. D. Chatterji, E. B. Dynkin, G. Kallianpur, D. Nualart, B. Øksendal, F. Proske, M. Röckner, and A. V. Skorokhod, for discussions which clearly have influenced the ideas presented in this book. In addition, V. Mandrekar thanks his colleague Y. Xiao, and his students J. Du and S. Zhang for continuously providing ideas in the early stage of development of the monograph.

We shall be remiss if we do not thank our wives Veena and Edyta for their immense patience and encouragement.

Acronyms

A^c	Complement of set A
S^ε	ε-neighbourhood of S
$S^{-\varepsilon}$	points in S whose distance from ∂S is greater than ε
$\mathcal{B} \vee \mathcal{G}$	The σ-field generated by $\mathcal{B} \cup \mathcal{G}$
$\mathbb{N}$	The set of positive integers
$d_c(I_H - A)$	Carleman-Fredholm Determinant of $A \in H^{\otimes 2}$
$C(t,t')$	Covariance
$(C_0(E), \|\cdot\|_\infty)$	The space of real-valued continuous functions with compact support in E with the supremum norm
$C([0,T],H)$	Banach space of H-valued continuous functions on $[0,T]$
$C^r([0,T],\mathbb{R})$	The space of real-valued r-times continuously differentiable functions on $[0,T]$
$C^r_{(0)}([0,T],\mathbb{R})$	The space of real-valued r-times continuously differentiable functions on $[0,T]$ vanishing at zero
$C^2_c(\mathbf{R}^n)$	The space of continuous functions on $\mathbf{R}^n$ with compact support and having bounded derivatives of the order up to two
$C^\infty_0(\mathbb{R}^n)$	The space of infinitely differentiable functions on $\mathbb{R}^n$ with compact support
$C^\infty_0(V)$	The space of infinitely differentiable functions on an open subset $V \subseteq \mathbb{R}^n$ with compact support
$D^{1,2}$	The domain of the Skorokhod derivative
$(\mathcal{F}_e, \mathcal{E})$	Regular extended (transient) Dirichlet space
GFMP	Germ Field Markov Property
$(H, \|\cdot\|_H)$	Hilbert space
$H(X)$	Gaussian space $\overline{\text{span}}\{X_t, t \in T\} \subseteq L^2(\Omega, \mathcal{F}, P)$ generated by Gaussian random field X
$H_1 \vee H_2$	$\overline{\text{span}}\{H_1, H_2\}, H_1, H_2 \subseteq H$
$\mathcal{F}_X$	σ-field $\sigma(H(X))$

l^2	The Hilbert space of square summable sequences
l^p	The Banach space of sequences summable with power p
S_n	Unit sphere in $\mathbb{R}^n$
L_n	Hyperbolic space in $\mathbb{R}^{n+1}$
λ	The Lebesgue measure
γ_n	The $N(0, I_{\mathbb{R}^n})$ distribution on $\mathbb{R}^n$
$M(E)$	The set of Radon signed measures with compact support in a separable locally compact Hasdorff space E
$M(D)$	The subspace of $K(C)$ defined as $\overline{\text{span}}\{f \mid f \in K(C) \text{and} \text{supp}(f) \subseteq D\}$
$\hat{M}(D)$	The subspace of $K(C)$ defined as $\overline{\text{span}}\{f_g \mid g \in G(E) \text{and} \text{supp}(g) \subseteq D\}$
$|\mu|$	Total variation measure of μ
$\text{supp}(\mu)$	Support of the measure μ
$L^p(\mathbb{R}^n)$	Banach space of real-valued functions on $\mathbb{R}^n$ integrable with power p with respect to the Lebesgue measure λ
$L^p(\mathcal{X}, d\mu)$	Banach space of real-valued functions on $\mathcal{X}^n$ integrable with power p with respect to the measure μ
$L^0(\Omega, \mathcal{F}, P)$	The vector space of real-valued measurable functions on $(\Omega, \mathcal{F})$ with the topology of convergence in probability.
$L^\infty(E, m)$	Banach space of real-valued essentially bounded functions on a measurable space (E, m)
$L^2(\mathbb{R}^n, \mathbb{C}, \lambda)$	Hilbert space of complex-valued square integrable functions on $\mathbb{R}^n$ with respect to the Lebesgue measure λ
$L^2(\Omega, K)$	Hilbert space of K-valued functions Bochner integrable in the second power
$W^{p,2}(\mathbb{R}^n)$	Sobolev space of functions $f \in L^p(\mathbb{R}^n)$, with $D^\alpha f \in L^p(\mathbb{R}^n)$, $|\alpha| \le p$
$W_0^{p,2}(V)$	Sobolev space of functions $f \in L^p(V)$ with compact support, with $D^\alpha f \in L^p(V)$, $|\alpha| \le p$
T^*	Adjoint of a linear operator T
$E_P(X)$	The integral $\int_\Omega X(\omega) \, dP(\omega)$
$E_{P_X}(f)$	The integral $\int_{\mathbb{R}^n} f(x) \, dP_X(x)$
$E^\mathcal{G} X$	Conditional expectation of a random variable X given a σ-field $\mathcal{G}$
$H^1(\mathcal{X} \to H, d\mu)$	Functions $G: \mathcal{X} \to H$, which are SGD and RAC
$\{h_n\}_{n=0}^\infty$	Normalized Hermite polynomials

$\{\tilde{h}_n\}_{n=0}^{\infty}$ Standard Hermite polynomials

$\{H_n\}_{n=0}^{\infty}$ Hermite polynomials

$H(X:S)$ The subspace $\overline{\mathrm{span}}\{X_\mu, \mathrm{supp}(\mu) \subseteq S\} \subseteq L^2(\Omega, \mathcal{F}, P)$

$\overline{H}(X:S)$ The subspace $\bigcap_{\mathcal{O} \supseteq S} \overline{\mathrm{span}}\{X_\mu, \mathrm{supp}(\mu) \subseteq \mathcal{O}\} \subseteq L^2(\Omega, \mathcal{F}, P)$, $\mathcal{O}$ are open

$K(C)$ The Reproducing Kernel Hilbert Space of a covariance C

$K(C)^{\otimes p}$ The pth tensor product of $K(C)$

$K(C)^{\odot p}$ The subspace of $K(C)^{\otimes p}$ consisting of all symmetric functions

$K(S)$ The subspace of $K(C)$ defined as $\pi^{-1}(H(X:S)) = \overline{\mathrm{span}}\{C(\cdot, \mu) \mid \mathrm{supp}(\mu) \subseteq S\}$

$\mathcal{L}(X, Y)$ Bounded linear operators from X to Y

$\mathcal{L}_2(H_1, H_2)$ Hilbert-Schmidt operators from H_1 to H_2

$\mathcal{L}_1(H_1, H_2)$ Trace-class operators from H_1 to H_2

∇F Malliavin derivative of $F \in L^2(\Omega)$

$D_t F$ Skorokhod derivative of $F \in L^2(\Omega)$

$\tilde{D}G$ Stochastic Gateaux derivative

LG Itô-Ramer integral

$\delta^e(g)$ Ogawa integral of a K-valued Bochner measurable function $g : \Omega \to K$ with respect to a particular orthonormal basis

$\delta^e(g)$ Universal Ogawa integral

$I^s(u)$ Skorokhod integral of $u \in L^2(\Omega, K)$

$(a_1 * a_2)(t)$ Convolution of two functions

$\mathcal{D}(L)$ Domain of the Itô-Ramer integral

$\mathcal{D}(f)$ Domain of the function f

SGD   Stochastic Gateaux differentiable

RAC   Ray absolutely continuous

Chapter 1

Covariances and Associated Reproducing Kernel Hilbert Spaces

The purpose of this chapter is to review facts about covariances and explain how they are related to Reproducing Kernel Hilbert Spaces (RKHSs). We provide fundamental facts about RKHSs and discuss typical examples relevant to the theory of Gaussian processes.

1.1 Covariances and Negative Definite Functions

In view of Kolmogorov consistency theorems (see [18]), a centered process can be constructed on any set T, once we have a function C on $T \times T \to \mathbb{R}$ satisfying the following conditions:

(SM) (Symmetry) For all $t, t' \in T$

$$C(t, t') = C(t', t). \tag{1.1}$$

(PD) (Positive Definiteness) For any finite number of distinct points $t_1, ..., t_n \in T$ and constants $u_1, u_2, ..., u_n \in \mathbb{R}$

$$\sum_{i,j=1}^{n} u_i u_j C(t_i, t_j) \geq 0. \tag{1.2}$$

We call $C(t, t')$ satisfying (SM) *symmetric* and satisfying (PD) *positive definite*.

Throughout the book we shall be using Hilbert spaces. For a Hilbert space H, we denote by $\langle \cdot, \cdot \rangle_H$ its inner product and by $\| \cdot \|_H$ its norm.

In order to construct examples of positive definite functions, we give equivalent conditions to (1.1) and (1.2).

Lemma 1.1.1. *Let T be a set and $C : T \times T \to \mathbb{R}$. Then the following are equivalent:*

(a) *There exists a real Hilbert space H and a transformation $f : T \to H$ such that*

$$C(t, t') = \langle f(t), f(t') \rangle_H \quad \text{for all } t, t' \in T.$$

(b) *There exists a family $\{f_j\}_{j \in I}$ (with I denoting an index set) of functions* *from T to $\mathbb{R}$, such that $\sum_{j\in I} f_j^2(t)$ is finite for each t and for $t,t' \in T$*

$$C(t,t') = \sum_{j\in I} f_j(t)f_j(t').$$

(c) *C satisfies* (1.1) *and* (1.2).

Proof. (a) $\Rightarrow$ (b). Let H be a real Hilbert space and $\{e_j\}_{j\in I}$ be its orthonormal basis. Then for $t \in T$,

$$f(t) = \sum_{j\in I} \langle f(t), e_j\rangle_H e_j.$$

Let $f_j(t) = \langle f(t), e_j\rangle_H$, for $j \in I$. By Parseval's identity $\sum_{j\in I} f_j^2(t) \leq \|f(t)\|_H^2$ for each t and for $t,t' \in T$

$$\langle f(t), f(t')\rangle_H = \sum_{j\in I} f_j(t)f_j(t').$$

To prove that (b) $\Rightarrow$ (a), we take any Hilbert space with cardinality of the basis $\{e_j\}_{j\in I}$ equal to cardinality of I. Since $\sum_{j\in I} f_j^2(t)$ is finite, $f(t) = \sum_{j\in I} f_j(t)e_j$ is well defined as a function from T to H. Now using Parseval's identity, we obtain

$$\langle f(t), f(t')\rangle_H = \sum_{j\in I} f_j(t)f_j(t').$$

(a) $\Rightarrow$ (c). C defined as in (a) is symmetric. Assume $t_1,...,t_n$ and $u_1,...,u_n$ are as in (1.2). Then, by bilinearity of the inner product

$$\sum_{i,j=1}^{n} u_iu_jC(t_i,t_j) = \sum_{i,j=1}^{n} u_iu_j\langle f(t_i),f(t_j)\rangle_H = \left\|\sum_{i=1}^{n} u_if(t_i)\right\|_H^2 \geq 0.$$

Now, it remains to prove (c) $\Rightarrow$ (a). This involves constructing a Hilbert space given C satisfying conditions (1.1) and (1.2). Let $\mathbb{R}^T$ be the real vector space of all real-valued functions on T and M be the linear manifold generated by $\{C_t(\cdot)\}_{t\in T}$, where $C_t(t') = C(t,t')$ for $t' \in T$. For any two elements in M,

$$f = \sum_{i=1}^{n} a_iC_{s_i}, \quad g = \sum_{j=1}^{m} b_jC_{t_j}(\cdot), \quad a_i,b_j \in \mathbb{R}, i = 1,...n, j = 1,...,m,$$

define

$$\langle f,g\rangle = \sum_{i,j} a_ib_jC(s_i,t_j). \tag{1.3}$$

The real valued function $\langle f,g\rangle$ of $f,g \in M$ does not depend on the particular representation of f and g as

$$\langle f,g\rangle = \sum_{i=1}^{n} a_ig(s_i) = \sum_{j=1}^{m} b_jf(t_j).$$

Using the fact that $C(s_i,t_j) = C(t_j,s_i)$ we conclude that $\langle f,g \rangle$ is a bilinear symmetric function on $M \times M$ satisfying $\langle f,C_t \rangle = f(t)$ for all $f \in M$ and $t \in T$. Hence, we have $|f(t)| \le \|f\| C^{1/2}(t,t)$ for all $f \in M$, with $\|f\| = \langle f,f \rangle^{1/2}$. Hence $\|f\| = 0$ if and only if $f(t) = 0$ for all $t \in T$. Let $\overline{M}$ be the completion of the normed linear space $(M, \|\cdot\|)$ and $\|\cdot\|_{\overline{M}}$ be the extension of $\|\cdot\|$ to $\overline{M}$. Then $\left(\overline{M}, \|\cdot\|_{\overline{M}}\right)$ is a Hilbert space and $C(t,t') = \langle C_t(\cdot), C_{t'}(\cdot) \rangle_{\overline{M}}$ giving (a). $\qquad\square$

Definition 1.1.1. *A function $C:T \times T \to \mathbb{R}$ satisfying conditions (1.1) and (1.2) is called a covariance.*

Exercise 1.1.1. **(a)** *Show that a product of covariances is a covariance. Hint: Use Lemma 1.1.1, part (b).*

(b) *Show that a finite sum of covariances is a covariance. Hint: Use Lemma 1.1.1, part (c).*

(c) *Show that a pointwise limit of covariances is a covariance.*

Corollary 1.1.1. *Let f be a function on $\mathbb{R}$ defined by $f(u) = \sum_{k=0}^{\infty} a_k u^k$, $u \in \mathbb{R}$ and $a_k \ge 0$ for all k. Let C be a covariance on T such that $\sum_{k=0}^{\infty} a_k (C(t,t'))^k$ converges for each $t,t' \in T$, then $f(C(t,t'))$ is a covariance.*

Exercise 1.1.2. *Show that $f(C(t,t'))$ is a covariance for*

(a) $f(u) = e^{au}$, $a > 0$,

(b) $f(u) = (1-u)^{-s}, s > 0$, if $|C(t,t')| < 1$,

(c) $f(u) = \arcsin(u)$, if $|C(t,t')| \le 1$.

Example 1.1.1. *We provide examples of covariances and demonstrate the use of Lemma 1.1.1 and Corollary 1.1.1.*

(a) *Let T be the set of positive integers and $C(t,t') = 0$ if $t \ne t'$ and $C(t,t') = \sigma_t^2 > 0$ for $t = t'$. Obviously, $C(t,t')$ is a covariance. We can also take*

$$H = l^2 = \left\{ x = (x_1, x_2, \ldots) \in \mathbb{R}^{\infty} \,\Big|\, \sum_{t=1}^{\infty} x_t^2 < \infty \right\}$$

and choose its ONB $e^t = (0,\ldots,0,1,0,\ldots)$, where $e_t^t = 1$. Then, defining $f(t) = \sigma_t e^t$ we relate C to the Hilbert space H and the function f as in (a) of Lemma 1.1.1 to give an alternative argument that C is a covariance on T.

(b) *Let $T = \{t = (t_1, \cdots, t_n) \in \mathbb{R}^n, t_i \ge 0, i = 1,2,\ldots,n\}$ and $C(t,t') = \prod_{i=1}^{n} t_i \wedge t_i'$, $t,t' \in T$. Then we can take $H = L^2(T,\lambda)$, the space of real-valued functions, square integrable with respect to the Lebesgue measure, and $f(t) = 1_{[0,t_1] \times \ldots \times [0,t_n]}(\cdot) \in H$. Using part (a) of Lemma 1.1.1, we can see that C is a covariance.*

(c) *Let $T = H$ be a Hilbert space and consider $C(t,t') = \langle t,t' \rangle_H$, then by the Riesz representation, for all t, we identify $\langle t, \cdot \rangle_H \in H^*$ with an element of H. Then we write $C(t,t') = \langle \langle t, \cdot \rangle_H, \langle t', \cdot \rangle_H \rangle_H$, so that $\langle t,t' \rangle_H$ is a covariance on T.*

(d) *Let $T = H$, then by Corollary 1.1.1, $C(t,t') = e^{a\langle t,t'\rangle_H}$ with $a > 0$ is a covariance on H.*

(e) *Let $T = S_n = \left\{ t = (t_1, ..., t_n) \in \mathbb{R}^n \mid \sum_{i=1}^n t_i^2 = 1 \right\}$, a unit sphere centered at the origin in $\mathbb{R}^n$, and $C(t,t') = \pi/2 - \Psi(t,t')$ where Ψ is the geodesic distance, that is, $\cos\Psi(t,t') = \sum_{i=1}^n t_i t_i' = t \cdot t'$, $0 \le \Psi(t,t') < \pi$. Then, since $t \cdot t'$ is a covariance we conclude by Corollary 1.1.1 that $C(t,t')$ is a covariance.*

(f) *Let $T = L_n = \left\{ t = (t_0, t_1, ..., t_n) \in \mathbb{R}^{n+1} \mid t_0^2 - \left(t_1^2 + ... + t_n^2 \right) = 1, t_0 > 0 \right\}$, a hyperbolic space of dimension n. Note that $t_0 > 0$ implies $t_0 > 1$. For $t,t' \in T$ define $[t,t'] = t_0 t_0' - t_1 t_1' - ... - t_n \cdot t_n'$. Let $C_0(t,t') = \left(t_0 t_0' \right)^\alpha$ for $\alpha \in \mathbb{R}$, then C_0 is a covariance on T by part (b) of Lemma 1.1.1. Also $C_1(t,t') = \sum_{i=1}^n t_i t_i'$ is a covariance on T. Hence, by Exercise 1.1.1, $C_2(t,t') = \left(t_0 t_0' \right)^{-1} C_1(t,t')$ is a covariance on T. Now*

$$\left| C_2(t,t') \right|^2 \le t_0^{-2} t_0'^{-2} \left(\sum_{i=1}^n t_i^2 \right) \left(\sum_{i=1}^n t_i'^2 \right),$$

so that

$$\left| C_2(t,t') \right|^2 \le t_0^{-2} t_0'^{-2} \left(t_0^2 - 1 \right) \left(t_0'^2 - 1 \right).$$

Hence $\left| C_2(t,t') \right| < 1$. Also, $1 - C_2(t,t') = \left(t_0 t_0' \right)^{-1} [t,t']$. Using Exercise 1.1.2 we obtain that $\left(t_0 t_0' \right)^s [t,t']^{-s}$ is a covariance, giving also that $[t,t']^{-s}$ is a covariance for $s > 0$.

(g) *Let $T = C_0^\infty(\mathbb{R}^n)$, the space of infinitely differentiable functions with compact support on $\mathbb{R}^n$. Let $C(t,t') = \int_{\mathbb{R}^n} tt' \, d\lambda$. With $H = L^2(\mathbb{R}^n, \lambda)$ we can see that $C(t,t')$ is a covariance on T.*

(h) *Let (S, Σ, μ) be a measurable space with a non-negative σ-finite measure μ. Let $T = \{ A \in \Sigma \mid \mu(A) < \infty \}$. Then, to see that $C(A,A') = \mu(A \cap A')$ is a covariance on T we consider $H = L^2(S, \Sigma, \mu)$ and $f(A) = 1_A(\cdot)$, for $A \in T$ and use part (a) of Lemma 1.1.1.*

Lemma 1.1.1 shows that conditions 1.1 and 1.2 on a function C on $T \times T$ are equivalent to the condition that $C(t,s)$ is an inner product of two values of some Hilbert-space valued function f. Schoenberg [3] asked when a symmetric function d defined on $T \times T$ satisfies the condition that $d(t,t') = \| g(t) - g(t') \|_H$ where g is a function from T to a Hilbert space H. In particular, this answers the question: if T is a metric space with distance d, then when is it isomorphic to a Hilbert space? As we are interested in the covariances, it is obvious that if d satisfies Schoenberg's condition, g is as above with $g(0) = 0$, $0 \in T$, then

$$\frac{1}{2} \left(d^2(t,0) + d^2(0,t') - d^2(t,t') \right)$$

is a covariance.

We now present Schoenberg's Theorem.

Theorem 1.1.1 (Schoenberg). *Let T be a set and d be a non-negative, symmetric real-valued function on $T \times T$ such that $d(t,t) = 0$. Then the following are equivalent,*

(a) *There exists a Hilbert space H and a function $g : T \to H$, such that for $t,t' \in T$*

$$d\left(t,t'\right) = \left\| g(t) - g(t') \right\|_H .$$

(b) *For any distinct points $t_1, t_2, ..., t_n$ in T and any real numbers $u_1, u_2, ..., u_n$ with $\sum_{i=1}^{n} u_i = 0$, with n an arbitrary positive integer,*

$$\sum_{i,j=1}^{n} d^2 \left(t_i, t_j \right) u_i u_j \leq 0.$$

(c) *For every real number $s > 0$, $e^{-sd^2(t,t')}$ is a covariance on T.*

Proof. (a) $\Rightarrow$ (c).

Let $C(t,t') = \langle g(t), g(t') \rangle_H$, then by Lemma 1.1.1 and Corollary 1.1.1, $e^{2sC(t,t')}$ is a covariance on T. Observe that for any positive integer n, $t_1, ..., t_n \in T$ and $a_1, ..., a_n \in \mathbb{R}$,

$$\sum_{i,j=1}^{n} e^{-sd^2(t_i,t_j)} a_i a_j = \sum_{i,j}^{n} e^{2sC(t_i,t_j)} b_i b_j$$

with $b_i = a_i e^{-s\|g(t_i)\|_H^2}$, $i = 1, ..., n$, proving (c).

(c) $\Rightarrow$ (b) For $s > 0$, and a positive integer n, let

$$\varphi(s) = \sum_{i,j=1}^{n} e^{-sd^2(t_i,t_j)} u_i u_j$$

for any $u_1, ..., u_n \in \mathbb{R}$ such that $\sum_{i=1}^{n} u_i = 0$, and with $t_1, ..., t_n \in T$. By (c), $\varphi(s) \geq 0$ and $\varphi(0) = 0$. Hence

$$\varphi'(0) = \lim_{s \downarrow 0} \frac{\varphi(s) - \varphi(0)}{s} \geq 0.$$

But $\varphi'(0) = -\sum_{i,j=1}^{n} d^2(t_i, t_j) u_i u_j$ giving (b).

(b) $\Rightarrow$ (a) Let 0 be an element of T. Consider $T_1 = T \setminus \{0\}$. Define

$$C(t,t') = \frac{1}{2} \left(d^2(0,t) + d^2(0,t') - d^2(t,t') \right) \quad \text{for } t,t' \in T. \tag{1.4}$$

Note that $C(0,t) = 0$ for all $t \in T$. For $t_1, ..., t_n \in T_1$ and $u_1, ..., u_n \in \mathbb{R}$ we have

$$2 \sum_{i,j=1}^{n} C(t_i,t_j) u_i u_j = - \sum_{i,j=0}^{n} d^2(t_i,t_j) u_i u_j \tag{1.5}$$

where $t_0 = 0$ and $u_0 = -(u_1 + \cdots + u_n)$. Since $\sum_{i=0}^n u_i = 0$, we conclude by (1.5) and (b) that C is a covariance on T_1. By Lemma 1.1.1 there exists a Hilbert space H and a function $f : T_1 \to H$, such that

$$C(t,t') = \langle f(t), f(t') \rangle_H \quad t,t' \in T_1.$$

Define g on T, by $g(0) = 0$ and $g(t) = f(t), t \in T_1$. Then

$$C(t,t') = \langle f(t), f(t') \rangle_H \quad \text{and} \quad d(t,t') = \|g(t) - g(t')\|_H, \quad t,t' \in T.$$

$\square$

Motivated by Schoenberg's theorem, we provide definition of a negative definite function.

Definition 1.1.2. *A symmetric function $\psi : T \times T \to \mathbb{R}$ is called negative definite if for any $t \in T$, $\psi(t,t) = 0$ and for $u_1,...,u_n \in \mathbb{R}$, such that $\sum_{i=1}^n u_i = 0$, and $t_1,...,t_n \in T$, where n is an arbitrary positive integer,*

$$\sum_{i,j=1}^n \psi(t_i,t_j) u_i u_j \le 0.$$

Exercise 1.1.3.

(a) *Let T be a set and ψ be a negative definite on $T \times T$. Let $0 \in T$. Show that the function*

$$C(t,t') = \frac{1}{2} \big(\psi(0,t) + \psi(0,t') - \psi(t,t') \big), \quad t,t' \in T,$$

is a covariance on T.

(b) *Let T be an additive group and $\varphi : T \to \mathbb{R}$ be symmetric, that is, $\varphi(t) = \varphi(-t)$, $t \in T$, and satisfy condition*

$$\sum_{i,j=1}^n \big(\varphi(t_i) + \varphi(t_j) - \varphi(t_i - t_j) \big) a_i a_j \ge 0$$

for any positive integer n, with $t_1,...,t_n \in T$ and $a_1,...,a_n \in \mathbb{R}$. Show that $\psi(t,t') = \varphi(t - t')$ is a negative definite function.

Corollary 1.1.2. *Let ψ be a negative definite function. Then*

(a) *For each $\alpha > 0$, $\psi_\alpha(t,t') = 1 - e^{-\alpha \psi(t,t')}$ is negative definite.*

(b) *Let f be a function on $\mathbb{R}_+$ given by*

$$f(u) = \int_0^\infty \frac{1}{\alpha}(1 - e^{-\alpha u}) \mu(d\alpha),$$

where μ is a non-negative measure on $\mathbb{R}_+$, such that

$$\int_0^\infty \frac{1}{1+\alpha} \mu(d\alpha) < \infty.$$

Then for any negative definite function ψ on T, the function $f(\psi)$ is negative definite.

Proof. In view of Definition 1.1.2, it is enough to prove (a). We know from Theorem 1.1.1 that

$$C_\alpha(t,s) = e^{-\alpha\psi(t,s)}, \quad C_\alpha(t,t) = 1$$

is a covariance on T. Hence there exists a Hilbert space H and $f_\alpha : T \to H$ such that $C_\alpha(t,t') = \langle f_\alpha(t), f_\alpha(t') \rangle_H$ for $t,t' \in T$. Hence for $t,t' \in T$,

$$\begin{aligned}
1 - e^{-\alpha\psi(t,t')} &= \frac{1}{2}\left(C_\alpha(t',t') + C_\alpha(t,t) - 2C_\alpha(t,t')\right) \\
&= \frac{1}{2}\left\|f_\alpha(t) - f_\alpha(t')\right\|_H^2
\end{aligned}$$

is negative definite by (b) of Theorem 1.1.1. □

Exercise 1.1.4.

(a) *Show that for $u > 0$,*

$$\ln(1+u) = \int_0^\infty \frac{1}{\alpha}\left(1 - e^{-\alpha u}\right)e^{-\alpha}\,d\alpha.$$

(b) *Show that for $0 < \beta < 1$ and $u > 0$,*

$$u^\beta = \frac{\beta}{\Gamma(1-\beta)} \int_0^\infty \frac{1}{\alpha}\left(1 - e^{-\alpha u}\right)\frac{1}{\alpha^\beta}\,d\alpha.$$

(c) *Let $T = H$ be a real Hilbert space. Show that $\psi(t,s) = \|t-s\|_H^\alpha$ for $0 < \alpha < 1$ is a negative definite function.*

(d) *Let $T = S_n$ be the sphere in $\mathbb{R}^{n+1}$ as in Example 1.1.1. Show that the geodesic distance on S_n is negative definite.*

(e) *Let $T = L_n$ as in Example 1.1.1 and for $t,t' \in L_n$ define*

$$[t,t'] = t_0 t_0' - \sum_{i=1}^{n} t_i t_i'.$$

Geodesic distance on L_n is defined by $\psi(t,t') = \cosh^{-1}[t,t']$. Show that ψ is negative definite.

Remark 1.1.1. *Part (c) of Exercise 1.1.4 is valid for $0 < \alpha \le 2$. Indeed, the function*

$$\varphi(t,s) = e^{-a\|t-s\|_H^\alpha}, \quad a > 0,\ t,s \in H$$

is a characteristic function of the increment $X_t - X_s$, where X_t is a $S\alpha S$ (symmetric α stable) process. Hence, $\varphi(\cdot,\cdot)$ is positive definite and obviously symmetric. Then it is a covariance, implying by Theorem 1.1.1 that $\|t-s\|_H^\alpha$ is negative definite for $0 < \alpha \le 2$.

1.2   Reproducing Kernel Hilbert Space

In the proof of Lemma 1.1 ((c) $\Rightarrow$ (a)) we constructed the completion H of a pre-Hilbert space M, a linear manifold generated by $\{C_t(\cdot),\ t \in T\}$. Recall that for $f \in M$, we had the reproducing property $\langle f, C_t(\cdot) \rangle = f(t)$. Hence, if $\{f_n\}_{n=1}^\infty \subseteq M$, is such that $f_n \to h \in H$, then the inequality

$$|\langle f_n - f_m, C_t(\cdot) \rangle| \le \|f_n - f_m\| C^{1/2}(t,t)$$

implies that for a fixed t,

$$\{\langle f_n, C_t(\cdot) \rangle\}_{n=1}^\infty$$

is a Cauchy sequence in $\mathbb{R}$, so that $\langle f_n, C_t \rangle$ converges to a limit. By combining two different sequences into one we can see that this limit does not depend on the choice of the sequence $f_n \to h$. We can thus define

$$f_h(t) = \langle h, C_t(\cdot) \rangle.$$

Then $h \to f_h(t)$ is a one-to-one map. Hence, we can define an inner product on the class of functions $\{f_h(t) : h \in H\}$ by

$$\langle f_h, f_g \rangle_1 = \langle h, g \rangle_H.$$

In conclusion, the space $\{f_h(t), h \in H\}$ is a completion of M consisting of functions. We denote this space by $K(C)$ and define $\langle \cdot, \cdot \rangle_{K(C)} = \langle \cdot, \cdot \rangle_1$. Thus, $K(C)$ is a Hilbert space possessing the following two properties:

(RKHS1) For each $t \in T$, $C_t(\cdot) \in K(C)$
(RKHS2) For every $t \in T$ and $f \in K(C)$, $\langle f, C_t(\cdot) \rangle_{K(C)} = f(t)$ (1.6)

The next theorem provides a result on the uniqueness of the space $K(C)$.

Theorem 1.2.1. *Given a covariance C on T, there exists exactly one Hilbert space $K(C)$ of functions on T satisfying (1.6). Furthermore, if $K(C) = K(C')$ for another covariance C' on T, then $C(t,t') = C'(t,t')$ for $t,t' \in T$.*

Proof. The first part follows from the construction of $K(C)$. If $K(C) = K(C')$ then $C_t'(\cdot) \in K(C)$ and therefore $\langle C_t'(\cdot), C_{t'}'(\cdot) \rangle_{K(C)} = C'(t,t')$. But since also $C_t(\cdot) \in K(C')$ we get $\langle C_{t'}(\cdot), C_{t'}(\cdot) \rangle_{K(C')} = C(t,t')$ giving the result.   □

Definition 1.2.1. *We call the unique Hilbert space $K(C)$ of functions f satisfying conditions (1.6) the Reproducing Kernel Hilbert Space of the covariance C on T.*

Example 1.2.1. *We now provide examples of RKHSs.*

(a) *Let $T = H$ be a Hilbert space and $C(t,t') = \langle t,t' \rangle_H$. Using the fact that H is isomorphic to its dual H^*, we obtain that H^* is an RKHS of H.*

(b) *Let* $T = \{t = (t_1, ..., t_n) \in \mathbb{R}^n, \ t_i \geq 0, \ i = 1, 2, ..., n\}$ *and* $C(t, t') = \prod_{i=1}^{n} \min(t_i, t_i')$ *for* $t, t' \in T$. *Consider*

$$K(C) = \left\{ f \mid f(t) = \int_0^{t_n} \cdots \int_0^{t_1} g_f(u_1, ..., u_n) \, du_1 ... du_n, \ g_f \in L^2(T, \lambda) \right\}.$$

Define

$$\langle f, h \rangle_{K(C)} = \langle g_f, g_h \rangle_{L^2(T, \lambda)}.$$

Since

$$C_t(t') = \int_0^{t_n'} \cdots \int_0^{t_1'} 1_{[0,t_1] \times ... \times [0,t_n]}(u_1, ..., u_n) du_1, ..., du_n$$

conditions (1.6) are satisfied for $K(C)$ *to be the RKHS of* C.

(c) *Let* $T = \mathbb{N}$ *and* $C(t, t') = \sigma_t^2$ *for* $t = t'$ *and* $C(t, t') = 0$ *for* $t \neq t'$. *Then* $K(C) = \{f \in l^2 \mid f(t) = \sigma_t g_f(t)\}$ *with the scalar product* $\langle f, h \rangle_{K(C)} = \sum_t \sigma_t^2 g_f(t) g_h(t)$.

(d) *Let* $T = C_0^\infty(\mathbb{R}^n)$, *the space of infinitely differentiable functions with compact support.*

(i) *If* $C(t, t') = \int_{\mathbb{R}^n} t(u) t'(u) \, du$. *Then* $K(C) = L^2(\mathbb{R}^n)$.

(ii) *If* $C(t, t') = \sum_{|\alpha| \leq p} \int_{\mathbb{R}^n} D^\alpha t D^\alpha t' \, d\lambda$ *where* $\alpha = (\alpha_1, ..., \alpha_n)$, $|\alpha| = \alpha_1 + ... + \alpha_n$, *and* $D^\alpha t = \frac{\partial^\alpha}{\partial^{\alpha_1} u_1 ... \partial^{\alpha_n} u_n} t(u_1, ..., u_n)$ *then* $K(C) = W^{p,2}(\mathbb{R}^n)$, *the Sobolev space of order* p.

(e) *Let* $T = \mathbb{R}^n$ *and* $C(t, s) = \frac{1}{2}(\|t\|_{\mathbb{R}^n} + \|s\|_{\mathbb{R}^n} - \|t - s\|_{\mathbb{R}^n})$, $t, s \in T$. *Then*

$$K(C) = \left\{ f \mid f(t) = Re \int_{\mathbb{R}^n} k_n^{-\frac{1}{2}} \|u\|_{\mathbb{R}^n}^{-\frac{n+1}{2}} \left(e^{itu} - 1\right) \overline{f_1(u)} \, du \right\}$$

where $f_1 \in L^2(\mathbb{R}^n, \mathbb{C}, \lambda)$, *with the scalar product*

$$\langle f, g \rangle_{K(C)} = Re \langle f_1, g_1 \rangle_{L^2(\mathbb{R}^n, \mathbb{C}, \lambda)}.$$

Here $L^2(\mathbb{R}^n, \mathbb{C}, \lambda)$ *denotes the space of complex-valued square-integrable functions with respect to the Lebesgue measure* λ *on* $\mathbb{R}^n$. *The constant* k_n *is given by*

$$k_n = \int_{\mathbb{R}^n} \frac{2(1 - \cos(e_1 \cdot \lambda))}{|\lambda|^{n+1}} \, d\lambda,$$

and $e_1 = (1, 0, ..., 0) \in \mathbb{R}^n$. *To see that this is the case, define* $f_t : \mathbb{R}^n \to \mathbb{C}$ *by*

$$f_t(u) = \|u\|_{\mathbb{R}^n}^{-\frac{n+1}{2}} \left(e^{it \cdot u} - 1\right), \quad t, u \neq 0,$$

and $f_0(u) = 0$. *Since for all* $u \in \mathbb{R}^n$

$$|f_t(u)|^2 = \frac{2(1 - \cos(t \cdot u))}{\|u\|_{\mathbb{R}^n}^{n+1}} \leq \min\left(\|t\|_{\mathbb{R}^n}^2 \|u\|_{\mathbb{R}^n}^{1-n}, 4\|u\|_{\mathbb{R}^n}^{-n-1}\right)$$

we can see that $f_t \in L^2(\mathbb{R}^n, \mathbb{C}, \lambda)$. Define $U : \mathbb{R}^n \to \mathbb{R}$ by $U(t) = \int_{\mathbb{R}^n} |f_t(u)|^2 du$, then $U(\alpha t) = \alpha U(t)$, for $\alpha > 0$, and $U(t)$ is invariant under orthogonal transformations of u and t, and continuous on $\mathbb{R}^n \setminus \{0\}$. Hence, $U(t) = k_n \|t\|_{\mathbb{R}^n}$, for $t \in \mathbb{R}^n$. Observe that $|f_t(u) - f_s(u)| = |f_{t-s}(u)|$, so that

$$\|f_t - f_s\|^2_{L^2(\mathbb{R}^n, \mathbb{C}, \lambda)} = U(t-s) = k_n \|t - s\|_{\mathbb{R}^n}.$$

We have that

$$C(t,s) = Re\left(k_n^{-1/2} f_t, k_n^{-1/2} f_s\right)_{L^2(\mathbb{R}^n, \mathbb{C}, \lambda)}$$

and since the set $\left\{k_n^{-1/2} f_t, t \in T\right\}$ generates $L^2(\mathbb{R}^n, \mathbb{C}, \lambda)$, the map $V : K(C) \to L^2(\mathbb{R}^n, \mathbb{C}, \lambda)$ defined by $V(C_t) = f_t$ is an isometry onto $L^2(\mathbb{R}^n, \mathbb{C}, \lambda)$. Now the result follows since for $f \in K(C)$,

$$f(t) = \langle C_t, f \rangle_{K(C)} = Re \langle V(C_t), V(f) \rangle_{L^2(\mathbb{R}^n, \mathbb{C}, \lambda)}.$$

(f) *Let (S, Σ, μ) be a measurable space with a σ-finite measure μ, $T = \{A \in \Sigma | \mu(A) < \infty\}$, and $C(A, A') = \mu(A \cap A')$, $A \in T$. Then $K(C) = \{v_f | v_f(A) = \int_A f d\mu, A \in T$ and $f \in L^2(S, \Sigma, \mu)\}$ with $\langle v_f, v_g \rangle_{K(C)} = \int f g d\mu$.*

Exercise 1.2.1. *Prove the claims made in (a) – (d) and (f) of Example 1.2.1.*

Remark 1.2.1. *Note that for $n = 1$ the covariances in parts (b) and (e) of Example 1.2.1 coincide.*

We now consider the tensor product of RKHSs. This material will be used in Chapter 3 to define a stochastic integral with respect to a Gaussian process. Let C_1, C_2 be two covariances on T_1 and T_2, respectively. Let $K(C_1), K(C_2)$ be their respective RKHSs and $\{e_\alpha^1, \alpha \in I_1\}$ and $\{e_\beta^2, \beta \in I_2\}$ be the orthonormal bases in $K(C_1), K(C_2)$, respectively. Then the series

$$\sum_{\alpha, \beta} a_{\alpha\beta} e_\alpha^1(t_1) e_\beta^2(t_2) \quad \text{with} \quad \sum_{\alpha, \beta} a_{\alpha, \beta}^2 < \infty \tag{1.7}$$

converges absolutely for each $(t_1, t_2) \in T_1 \times T_2$ (we leave this as Exercise 1.2.2). Hence, the series $\sum_{\alpha, \beta} a_{\alpha, \beta} e_\alpha^1(t_1) e_\beta^2(t_2)$ defines a function on $T_1 \times T_2$.

Exercise 1.2.2. *Prove the claim about the absolute convergence of the series (1.7).*

Consider

$$(C_1 \otimes C_2)((t_1, t_2), (t_1', t_2')) = C_1(t_1, t_1') C_2(t_2, t_2').$$

Then by part (b) of Exercise 1.1.1 $C_1 \otimes C_2$ is a covariance, which we call the tensor product of covariances C_1 and C_2.

Theorem 1.2.2. *The RKHS of the tensor product of two covariances C_1 and C_2 has the form*

$$K(C_1 \otimes C_2) = \left\{ f \mid f(t_1,t_2) = \sum_{\alpha,\beta} a_{\alpha\beta} e_\alpha^1(t_1) e_\beta^2(t_2), \ \sum_{\alpha,\beta} a_{\alpha,\beta}^2 < \infty \right\} \qquad (1.8)$$

and the scalar product is defined by

$$\langle f,g \rangle_{K(C_1 \otimes C_2)} = \sum_{\alpha,\beta} a_{\alpha,\beta} a_{\alpha,\beta}' \qquad (1.9)$$

for $f,g \in K(C_1 \otimes C_2)$ with

$$f(t_1,t_2) = \sum_{\alpha,\beta} a_{\alpha,\beta} e_\alpha^1(t_1) e_\beta^2(t_2) \quad and \quad g(t_1,t_2) = \sum_{\alpha,\beta} a_{\alpha,\beta}' e_\alpha^1(t_1) e_\beta^2(t_2).$$

Proof. Since for $i = 1,2$,

$$C_i(\cdot,t_i) = \sum_\alpha \langle C(\cdot,t_i), e_\alpha^i(\cdot) \rangle_{K(C_i)} e_\alpha^i(t_i) = \sum_\alpha e_\alpha^i(t_i) e_\alpha^i(\cdot)$$

with $\sum_\alpha (e_\alpha^i(t_i))^2 < \infty$, we have the following form of the tensor product of covariances,

$$(C_1 \otimes C_2)_{(\cdot,*)}(t_1,t_2) = \sum e_\alpha^1(t_1) e_\beta^2(t_2) e_\alpha^1(\cdot) e_\beta^2(*)$$

with $\sum_{\alpha,\beta} \left(e_\alpha^1(t_1) e_\beta^2(t_2) \right)^2 < \infty$. Using this representation, one can directly verify (1.6). □

Exercise 1.2.3. *Argue that $K(C_1 \otimes C_2) = K(C_1) \otimes K(C_2)$. Use induction to define $K(C_1 \otimes C_2 \otimes \cdots \otimes C_n)$.*

Example 1.2.2. *Let $T = \{ t = (t_1,...,t_n) \mid t_i \geq 0, \ i = 1,...,n \}$ and $C(t,t') = \prod_{i=1}^n \min(t_i,t_i')$, then $K(C) = K\left(C_1^{\otimes n} \right) = (K(C_1))^{\otimes n}$, where*

$$K(C_1) = \left\{ f \mid f(t) = \int_0^t g_f(u)\,du, \ g_f \in L^2(\mathbb{R}_+,\lambda) \right\}.$$

Exercise 1.2.4. *Describe the RKHSs in part (d) Example 1.2.1 as tensor products.*

We now consider bounded linear operators on RKHSs. Let $L : K(C_1) \to K(C_2)$ be a bounded linear operator, then $\Lambda(s,t) = L^* C_2(s,t)$, where L^* is the adjoint operator, satisfies $(Lf)(t) = \langle f(\cdot), \Lambda(\cdot,t) \rangle_{K(C_1)}$ for all $t \in T_2$. Conversely a function Λ on $T_1 \times T_2$ satisfying conditions

(a) $\Lambda(\cdot,t_2) \in K(C_1)$ for each $t_2 \in T_2$,

(b) for $g \in K(C_1)$, $\langle g(\cdot), \Lambda(\cdot,t_2) \rangle_{K(C_1)} \in K(C_2)$, as a function of $t_2 \in T_2$,

(c) $\sup_{t_2 \in T_2} \|\Lambda(\cdot, t_2)\|_{K(C_1)} < \infty$,

defines a bounded linear operator on $K(C_1)$ into $K(C_2)$. In particular, elements of $K(C_1 \otimes C_2)$ define bounded linear operators on $K(C_1)$ into $K(C_2)$. In fact, if $f \in K(C_1 \otimes C_2)$ then $(L_f g)(t_2) = \langle g, f(\cdot, t_2) \rangle_{K(C_1)}$.

Lemma 1.2.1. *Let $C_1 \otimes C_2$ be a covariance on $T_1 \times T_2$. Given $f \in K(C_1 \otimes C_2)$, $L_f : K(C_1) \to K(C_2)$ defined by*

$$(L_f g)(t_2) = \langle g, f(\cdot, t_2) \rangle_{K(C_1)}$$

is a Hilbert Schmidt operator. Conversely, for every Hilbert Schmidt operator $L : K(C_1) \to K(C_2)$ there exists a unique function $f \in K(C_1 \otimes C_2)$ such that $L = L_f$. In addition, $\|f\|_{K(C_1 \otimes C_2)}$ equals Hilbert-Schmidt norm of L_f.

Proof. Let $\{e_\alpha^1, \ \alpha \in I_1\}$ and $\{e_\beta^2, \ \beta \in I_2\}$ be orthonormal bases in $K(C_1)$ and $K(C_2)$, respectively, and $f(t_1, t_2) = \sum_\alpha \sum_\beta a_{\alpha\beta} e_\alpha^1(t_1) e_\beta^2(t_2)$. Then

$$(L_f g)(t_2) = \sum_\beta \left(\sum_\alpha a_{\alpha\beta} \langle g, e_\alpha^1 \rangle_{K(C_1)} \right) e_\beta^2(t_2).$$

Since

$$\|L_f g\|_{K(C_2)}^2 = \sum_\beta \left(\sum_\alpha a_{\alpha\beta} \langle g, e_\alpha^1 \rangle_{K(C_1)} \right)^2 \le \sum_{\beta,\alpha} a_{\alpha,\beta}^2 \|g\|_{K(C_1)}^2$$

the operator L is bounded. Since $(L_f e_\alpha^1)(t_2) = \sum_\beta a_{\alpha\beta} e_\beta(t_2)$, we get

$$\sum_\alpha \|L_f e_\alpha^1\|_{K(C_2)}^2 = \sum_{\alpha,\beta} a_{\alpha,\beta}^2$$

proving that L is a Hilbert–Schmidt operator.

If $L : K(C_1) \to K(C_2)$ is a Hilbert–Schmidt operator, then we define a function f on $T_1 \times T_2$ by

$$f(t_1, t_2) = (L^* C_2(t_2, \cdot))(t_1),$$

so that for $g \in K(C_1)$,

$$(Lg)(t_2) = \langle (Lg)(\cdot), C_2(t_2, \cdot) \rangle_{K(C_2)} = \langle g(\cdot), f(\cdot, t_2) \rangle_{K(C_1)}.$$

But $f(\cdot, t_2) \in K(C_1)$, and we have the following expansion:

$$f(t_1, t_2) = \sum_\alpha \langle f(\cdot, t_2), e_\alpha^1(\cdot) \rangle_{K(C_1)} e_\alpha^1(t_1).$$

Define

$$b_\alpha(t_2) = \langle L e_\alpha^1, C_2(\cdot, t_2) \rangle_{K(C_2)} = (L e_\alpha^1)(t_2) = \langle f(\cdot, t_2), e_\alpha^1(\cdot) \rangle_{K(C_1)},$$

Hence, $b_\alpha \in K(C_2)$ and

$$f(t_1,t_2) = \sum_\alpha \langle b_\alpha(t_2) e_\alpha^1(t_1) = \sum_{\alpha,\beta} a_{\alpha\beta} e_\beta^2(t_2) e_\alpha^1(t_1)$$

with $b_\alpha(t_2) = \sum_\beta a_{\alpha\beta} e_\beta^2(t_2)$. Using the assumption that L is a Hilbert–Schmidt operator, we conclude that

$$\|f\|_{K(C_1 \otimes K(C_2))} = \sum_\alpha \sum_\beta a_{\alpha,\beta}^2 = \sum_\alpha \|b_\alpha\|_{K(C_2)}^2 = \sum_\alpha \|Le_\alpha^1)\|_{K(C_2)}^2 < \infty$$

ensuring that $f \in K(C_1 \otimes C_2)$ and the equality of norms. $\square$

Next, we will identify the RKHS corresponding to a sum of two covariances. The following is a well-known theorem and we refer the reader to [3] for its proof.

Theorem 1.2.3. *Let $C_i(t,s)$, $t,s \in T$, $i = 1,2$ be two covariances with the corresponding RKHSs $(K(C_i), \|\cdot\|_i)$, $i = 1,2$. Then the RKHS $(K(C), \|\cdot\|_{K(C)})$ of $C(t,s) = C_1(t,s) + C_2(t,s)$ consists of all functions $f = f_1 + f_2$, with $f_i \in K(C_i)$, $i = 1,2$, and*

$$\|f\|_{K(C)}^2 = \inf\left\{ \|f_1\|_{K(C_1)}^2 + \|f_2\|_{K(C_2)}^2 \right\}$$

the infimum taken for all the decompositions $f = f_1 + f_2$ with $f_i \in K(C_i)$, $i = 1,2$.

Definition 1.2.2. *For two covariances we say that C_2 dominates C_1 and write $C_1 \ll C_2$ if $C_2 - C_1$ is a covariance.*

Exercise 1.2.5. *Show that if $C_1 \ll C_2$, then $K(C_1) \subseteq K(C_2)$.*

Hint: $C_2 = C_1 + (C_2 - C_1)$ implies that $f \in K(C_2)$ can be represented as $f_1 + f_2$, with $f_1 \in K(C_1)$ and $f_2 \in K(C_2 - C_1)$. In addition $\|f\|_{K(C_2)}^2 = \inf\{\|f_1\|_{K(C_1)}^2 + \|f_2\|_{K(C_2)}^2\}$, where the infimum is taken over all representations of f.

Let us give a more precise result than that in Exercise 1.2.5.

Lemma 1.2.2. *Let C_1 and C_2 be two covariances on T. Then $C_1 \ll C_2$ if and only if there exists a Hilbert space $\tilde{H}$ and mappings $f_i : T \to \tilde{H}$, $(i = 1,2)$ such that*

(a) *$\langle f_i(s), f_i(t)\rangle_{\tilde{H}} = C_i(s,t)$, $i = 1,2$.*

(b) *$f_1(t) = P(f_2(t))$, $t \in T$ where P is an orthogonal projection of $\tilde{H}$ on to the closed linear subspace $\overline{\text{span}}\{f_1(t), t \in T\}$ in $\tilde{H}$.*

Proof. The proof of the "if" part is left as an exercise (Exercise 1.2.6). Suppose $C_1 \ll C_2$. Let $K(C_1)$ and $K(C_3)$ be RKHSs of C_1 and $C_3 = C_2 - C_1$, respectively. Then for $j = 1,3$, $C_{j,t}(\cdot) \in K(C_j)$ and $\langle C_{j,s}, C_{j,t}\rangle_{K(C_j)} = C_j(s,t)$, $s,t \in T$.

Define the Hilbert space $\tilde{H}$ by

$$\tilde{H} = K(C_1) \oplus K(C_3) = \{(x,y)| x \in K(C_1), y \in K(C_3)\}$$

with the scalar product

$$\langle (x,y),(x',y') \rangle_{\tilde{H}} = \langle x,x' \rangle_{K(C_1)} + \langle y,y' \rangle_{K(C_3)}.$$

Let

$$f_1(t) = (C_{1,t},0), \quad f_2(t) = (C_{1,t},C_{3,t}), \quad t \in T.$$

Then for $t,s \in T$

$$\begin{aligned}
\langle f_1(t),f_1(s) \rangle_{\tilde{H}} &= C_1(t,s), \\
\langle f_2(t),f_2(s) \rangle_{\tilde{H}} &= C_1(t,s) + C_3(t,s) = C_2(t,s).
\end{aligned}$$

Also, $\overline{\text{span}}\{f_1(t), t \in T\} = K(C_1) \oplus \{0\}$ and hence

$$Pf_2(t) = (C_{1,t},0) = f_1(t), \quad t \in T.$$

$\square$

Exercise 1.2.6. *Prove the "if" part of Lemma 1.2.2.*
    Hint: Note that $\langle f_2(s),f_2(t) \rangle_{\tilde{H}} - \langle Pf_2(s),Pf_2(t) \rangle_{\tilde{H}} = \langle f_2(s)-Pf_2(s),f_2(t) - Pf_2(t) \rangle_{\tilde{H}}.$

Exercise 1.2.7. *Let* $L : K(C) \to K(C)$ *be a bounded linear operator and* $\Lambda(t',t) = L^*C(\cdot,t)(t')$. *Then*

(a) $\Lambda(\cdot,t) \in K(C)$ *and for* $f \in K(C)$, $(Lf)(t) = \langle f,\Lambda(\cdot,t) \rangle_{K(C)}.$

(b) *L is self-adjoint if and only if* Λ *is a symmetric function.*

(c) *L is a non-negative definite operator if and only if* Λ *is a covariance and there exists a constant* $k > 0$ *such that* $\Lambda \ll kC$.

(d) *L is a non-negative definite operator with a bounded inverse if and only if there exist constants* $0 < k_1 \le k_2$ *such that* $k_1 C \ll \Lambda \ll k_2 C$.

Exercise 1.2.8. *Let* T *be any set and* $\{T_\alpha, \alpha \in I\}$ *be a family of subsets of* T *such that* $T = \bigcup_\alpha T_\alpha$. *Let* C *be a covariance on* T. *Then* $f \in K(C)$ *if and only if* $f_\alpha = f|_{T_\alpha} \in K(C^{T_\alpha})$ *(restriction of* f *to* T_α*) and* $\sup_\alpha \|f_\alpha\|_{C^{T_\alpha}}$ *is finite, where* C^{T_α} *is the restriction of* C *to* $T_\alpha \times T_\alpha$.

Chapter 2

Gaussian Random Fields

2.1 Gaussian Random Variable

We begin with some properties of Gaussian random variables.

Definition 2.1.1. *Let $(\Omega, \mathcal{F}, P)$ be a probability space. A transformation X from $(\Omega, \mathcal{F})$ to a measurable space $(\mathcal{X}, \mathcal{A})$ is called an $\mathcal{X}$-valued random variable if $X^{-1}(A) \in \mathcal{F}$ for all $A \in \mathcal{A}$. In case $\mathcal{X} = \mathbb{R}^n$, the Euclidean space and $\mathcal{A} = \mathcal{B}(\mathbb{R}^n)$, then X will be referred to as an n-random vector. For $n = 1$, we call X a real-valued random variable (or just a random variable). A vector $X = (X_1, ..., X_n)$ is an n-random vector if and only if $X_1, X_2, \cdots, X_n$ are random variables.*

The probability measure $P_X = P \circ X^{-1}$ induced by X on $\mathcal{A}$ is called the distribution of X. The integral $\int_\Omega X(\omega) dP(\omega)$ will be written as $E_P(X)$, and the subscript P will be omitted if this does not lead to any confusion. Also, for a Borel measurable function $f : \mathbb{R}^n \to \mathbb{R}$ the integral $\int_{\mathbb{R}^n} f(x) dP_X(x)$ will be written as $E_{P_X}(f)$. For an n-vector X, the characteristic function of X is defined to be the function $\varphi_X : \mathbb{R}^n \to \mathbb{R}$, defined for all $t \in \mathbb{R}^n$, by

$$\varphi_X(t) = E_P e^{it \cdot X} = \int_{\mathbb{R}^n} e^{it \cdot u} P_X(du),$$

where above, for short, $u \cdot v$ denotes the inner product of u and v in $\mathbb{R}^n$ and $i = \sqrt{-1}$.

The following are well-known properties of characteristic functions [20]. Characteristic function φ_X uniquely determines the distribution P_X of a random variable X, that is, if $\varphi_X = \varphi_Y$, then $P_X = P_Y$. The coordinates of an n-random vector $X = (X_1, ..., X_n)$ are mutually independent if and only if $\varphi_X(t) = \prod_1^n \varphi_{X_i}(t_i)$, for all $t = (t_1, ..., t_n) \in \mathbb{R}^n$.

Definition 2.1.2. *A random variable X_σ defined on a probability space $(\Omega, \mathcal{F}, P)$ is called centered Gaussian with variance σ^2 if P_{X_σ} is given by*

$$P_{X_\sigma}(A) = \frac{1}{\sigma \sqrt{2\pi}} \int_A e^{-\frac{u^2}{2\sigma^2}} du \quad \text{for } A \in \mathcal{B}(\mathbb{R}). \tag{2.1}$$

The characteristic function of X_σ is given by

$$\varphi_{X_\sigma}(t) = e^{-\frac{t^2 \sigma^2}{2}} \quad \text{for } t \in \mathbb{R}. \tag{2.2}$$

Exercise 2.1.1. *Show that*

(a) $\left(\int_{\mathbb{R}} e^{-\frac{u^2}{2}} du \right)^2 = 2\pi.$

(b) $E|X_\sigma|^p = C_p \sigma^p$ *where* $C_p = \sqrt{2\pi} 2^{p/2} \Gamma\left(\frac{p+1}{2}\right).$

Let us denote by $\mu \equiv \nu$ the mutual absolute continuity of two measures μ and ν. Then we leave the following to be proved by the reader in Exercise 2.1.3.

Proposition 2.1.1. *Let X be centered Gaussian random variable with unit variance. Then*

(a) $P_X \equiv P_{X+m}$ *for* $m \in \mathbb{R}.$

(b) *The density of the distribution P_{X+m} with respect to P_X is given by*

$$\frac{dP_{X+m}}{dP_X}(u) = e^{mu - \frac{1}{2}m^2} \tag{2.3}$$

for $u \in \mathbb{R}.$

(c) *Denote by $d_m(u)$ the density in (2.3), then $d_m \in L^2(\mathbb{R}, P_X)$ and*

$$E_{P_X}(d_m d_{m'}) = e^{mm'}$$

for $m, m' \in \mathbb{R}.$

(d) *Let $f \in L^2(\mathbb{R}, P_X)$, then f has the following expansion*

$$f(u) = \sum_{n=0}^{\infty} E_{P_X}(f h_n) h_n(u) \tag{2.4}$$

where $\{h_n\}_{n=0}^{\infty}$ are Hermite polynomials [1],

$$h_n(u) = \frac{(-1)^n}{\sqrt{n!}} e^{\frac{u^2}{2}} \left(\frac{\partial^n}{\partial u^n} e^{-\frac{u^2}{2} + ut} \right)\Big|_{t=0} \tag{2.5}$$

and the series converges in $L^2(\mathbb{R}, P_X)$.

(e) *The densities $\{d_m, m \in \mathbb{R}\}$ generate the space $L^2(\mathbb{R}, P_X)$.*

Exercise 2.1.2. *Examine the following properties of Hermite polynomials.*

(a) *Show that $(\sqrt{n!}) h_n(u) = 2^{-n/2} \tilde{h}_n(u/\sqrt{2})$, where*

$$\tilde{h}_n(u) = \left(\frac{\partial^n}{\partial u^n} e^{-u^2 + 2ut} \right)\Big|_{u=0},$$

and $\{\tilde{h}(u)\}_{n=0}^{\infty}$ commonly define Hermite polynomials. The polynomials $h_n(u)$ are normalized versions of $\tilde{h}_n(u)$.

(b) *Show that the system $\{h_n\}_{n=0}^{\infty}$ is orthonormal in $L^2(\mathbb{R})$ with the weight*

$$\frac{1}{\sqrt{2\pi}} e^{-\frac{x^2}{2}}$$

Exercise 2.1.3. *Prove Proposition 2.1.1.*

Exercise 2.1.4. *State and prove an analogue of Proposition 2.1.1 for X_σ.*

If X_σ is centered Gaussian random variable, then $E(X_\sigma + m) = m$. We say that X is a Gaussian random variable with mean $m \in \mathbb{R}$, and variance σ^2 if $X - m$ is a centered Gaussian random variable with variance σ. Observe that

$$\varphi_{X_\sigma + m}(t) = e^{-\frac{1}{2}t^2\sigma^2 + imt} \quad \text{for } t \in \mathbb{R}.$$

Lemma 2.1.1. *Limit in probability of a sequence of (equivalence classes) of Gaussian random variables is itself a Gaussian variable and this convergence is also in L^p-norm for $1 \le p < \infty$. The limiting random variable is centered if each element of the sequence is centered.*

Proof. Since for $t \in \mathbb{R}$, $|t| \le K$,

$$\left| e^{itu} - e^{itv} \right| \le \min(2, K|u - v|)$$

we obtain that

$$\sup_{|t| \le K} \left| E e^{itX_n} - E e^{itX} \right| \le K\varepsilon + P\left(|X_n - X| > \varepsilon \right)$$

for all $\varepsilon > 0$. Hence, with $\varphi_{X_n}(t) = e^{-\frac{1}{2}\sigma_n^2 t^2 + im_n t}$ for all $t \in \mathbb{R}$,

$$E\left(e^{itX} \right) = \lim_{n \to \infty} e^{-\frac{1}{2}\sigma_n^2 t^2 + im_n t}$$

and the convergence in uniform on compact subsets of $\mathbb{R}$. This implies that σ_n^2 and m_n are convergent sequences. Let $\sigma^2 = \lim_n \sigma_n^2$ and $m = \lim_n m_n$, then

$$E\left(e^{itX} \right) = e^{-\frac{1}{2}\sigma^2 t^2 + imt}.$$

Since the sequence $e^{\frac{1}{2}\sigma_n^2}$ is bounded, we have

$$\sup_n E\left(e^{X_n} + e^{-X_n} \right) < \infty$$

proving uniform integrability of the sequence $\{|X_n|^p\}_{n=1}^\infty$ for $1 \le p < \infty$, which implies convergence in L^p by the assumption of the convergence in probability of X_n to X. □

Throughout, $G(m, \sigma^2)$ will denote the Gaussian random variable $X_\sigma + m$.

We note that if we choose i.i.d. random variables $X_n = G(m, \sigma^2)$, then the convergence in distribution to $X = G(m, \sigma^2)$ is obvious but $X_n \not\to X$ in L^2. Thus, the conclusion of Lemma 2.1.1 fails if convergence in probability is replaced by convergence in distribution.

We conclude this section with the following proposition, whose proof is left to the reader.

Proposition 2.1.2. *Let $X = (X_1, ..., X_n)$ be an n-random vector, then the following are true.*

(a) *The scalar product $u \cdot X$ is a random variable for all $u \in \mathbb{R}^n$.*

(b) *If $X_1, ..., X_n$ are independent Gaussian random variables, then $u \cdot X$ is a Gaussian random variable.*

(c) *If $u \cdot X$ is a centered Gaussian r.v. for each $u \in \mathbb{R}^n$ and $EX_iX_j = 0$, $i \neq j$, then $X_1, ..., X_n$ is a sequence of independent centered Gaussian random variables.*

Exercise 2.1.5. *Prove Proposition 2.1.2.*

2.2   Gaussian Spaces

We shall now introduce the definition of Gaussian random fields and their properties. Following Kakutani [51], we have the following definition.

Definition 2.2.1. *Let $(\Omega, \mathcal{F}, P)$ be a probability space, then a linear subspace $M \subseteq L^2(\Omega, \mathcal{F}, P)$ is called a Gaussian manifold if each $X \in M$ is a centered Gaussian random variable. A Gaussian manifold which is closed in $L^2(\Omega, \mathcal{F}, P)$ will be called a Gaussian space.*

Proposition 2.2.1.

(a) *The closure of a Gaussian manifold in $L^2(\Omega, \mathcal{F}, P)$ is Gaussian space.*

(b) *If $X_1, X_2, ..., X_n$ are orthogonal elements of a Gaussian manifold, then $X_1, X_2, ..., X_n$ are independent random variables.*

(c) *The closed linear subspace $H(X) = \overline{\text{span}}\{X_i, i \in I\}$ of $L^2(\Omega, \mathcal{F}, P)$ generated by a family of independent Gaussian random variables $\{X_i, i \in I\}$ is a Gaussian space.*

(d) *Let K be a Hilbert space. Then there exists a Gaussian space $H \subseteq L^2(\Omega, \mathcal{F}, P)$ isomorphic to K for some probability space $(\Omega, \mathcal{F}, P)$.*

Proof. The statements (a) through (c) follow from Lemma 2.1.1 and Proposition 2.1.2. To prove (d), choose an orthonormal basis $\{e_i, i \in I\}$ in K. Let $\Omega_i = \mathbb{R}$ and $\mathcal{F}_i = \mathcal{B}(\mathbb{R})$ and $P_i = G(0, 1)$ for all $i \in I$. The suitable probability space $(\Omega, \mathcal{F}, P)$ is the product probability space $(\prod_i \Omega_i, \otimes_i \mathcal{F}_i, \prod_i P_i)$. Denote by $X_i(w) = w_i$ and for $k \in K$, with $k = \sum_{i \in I} \langle k, e_i \rangle_K e_i$, define

$$\pi(k) = \sum_{i \in I} \langle k, e_i \rangle_K X_i. \tag{2.6}$$

Note that because $\sum_{i \in I} \langle k, e_i \rangle_K^2 = \|k\|_K^2 < \infty$, the series in (2.6) converges in $L^2(\Omega, \mathcal{F}, P)$ and

$$\langle k, k' \rangle_K = \sum_{i=1}^{\infty} \langle k, e_i \rangle_K \langle k', e_i \rangle_K = \langle \pi(k), \pi(k') \rangle_{L^2(\Omega, \mathcal{F}, P)}.$$

Hence, $H = \{\pi(k),\, k \in K\} \subseteq L^2(\Omega,\mathcal{F},P)$ is a Gaussian space isomorphic to K. □

We now present some properties of Gaussian subspaces of a Gaussian space H contained in $L^2(\Omega,\mathcal{F},P)$. For a random variable $X \in L^1(\Omega,\mathcal{F},P)$ and a σ-field $\mathcal{G} \subseteq \mathcal{F}$, we denote by $E^{\mathcal{G}}X$ the *conditional expectation* of X given $\mathcal{G}$.

Proposition 2.2.2.

(a) *Let $\{H_i,\, i \in I\}$ be a family of Gaussian subspaces of a Gaussian space H in $L^2(\Omega,\mathcal{F},P)$. Then the family of σ-fields $\sigma(H_i)$, generated by H_i, $i \in I$ are independent if and only if $H_j \perp H_k$ for $j \neq k$.*

(b) *Let H_1 and H_2 be two orthogonal subspaces of a Gaussian space H in $L^2(\Omega,\mathcal{F},P)$, then $E^{\sigma(H_1)}Z = EZ$ for all $Z \in L^2(\Omega,\sigma(H_2),P)$.*

(c) *Let H_1 be a subspace of a Gaussian space H and $X \in H$, then $E^{\sigma(H_1)}X = P_{H_1}X$, where P_{H_1} denotes orthogonal projection of H on H_1.*

Proof. The proofs of (a) and (b) are left to the reader as an exercise. To show (c), observe that $H = H_1 \oplus H_1^\perp$ is a decomposition of H and $X = X_1 + X_2$, with $X_1 \in H_1$ and $X_2 \in H_1^\perp$. Then $E^{\sigma(H_1)}X = X_1 + E(X_2)$. But $EX_2 = 0$, giving the result. □

Exercise 2.2.1. *Prove parts (a) and (b) of Proposition 2.2.2.*

Definition 2.2.2. *Let T be a set and $(\Omega,\mathcal{F},P)$ be a probability space. A family of random variables $\{X_t, t \in T\} \subseteq L^2(\Omega,\mathcal{F},P)$ is called*

(a) *A centered Gaussian random field if the real linear manifold generated by $\{X_t, t \in T\}$ is a Gaussian manifold.*

(b) *Gaussian random field with mean m if there exists an $m \in \mathbb{R}^T$ such that $\{X_t - m(t),\, t \in T\}$ is a centered Gaussian random field.*

Lemma 2.2.1. *Let $\{X_t, t \in T\}$ be a centered Gaussian random field on a probability space $(\Omega,\mathcal{F},P)$. Then*

(a) *For each $t \in T$, X_t is a centered Gaussian random variable.*

(b) *The subspace $H(X) = \overline{\operatorname{span}}\{X_t, t \in T\}$ of $L^2(\Omega,\mathcal{F},P)$ is a Gaussian space.*

(c) *The function $C_X(t,t') = \langle X_t, X_{t'} \rangle_{L^2(\Omega,\mathcal{F},P)}$ is a covariance on T.*

(d) *$\sigma(H(X)) = \sigma\{\sigma\{X_t, t \in T\} \cup \mathcal{N}\}$ where $\mathcal{N}$ is a class of sets of P-measure zero in $\mathcal{F}$.*

(e) *$K(C_X) = \{f : f(t) = EX_tY_f \text{ for a unique } Y_f \in H(X)\}$ with $\langle f,g \rangle_{K(C_X)} = (Y_f, Y_g)_{L^2(\Omega,\mathcal{F},P)}$. The map $V(f) = Y_f$ is an isometry of $K(C_X)$ onto $H(X)$.*

(f) *For each $t \in T$, $X_t = \sum_{i \in I} e_i(t)\xi_i$ where $\{\xi_i, i \in I\}$ are independent random variables with $e_i = V^{-1}\xi_i$ for all $i \in I$. For each $t \in T$, the series converges P–a.e. and in $L^2(\Omega,\mathcal{F},P)$.*

Proof. Parts (a) and (b) follow directly from Definitions 2.2.1, 2.2.2 and Proposition 2.2.1. Part (c) is a consequence of Lemma 1.1.1 and (d) follows from Lemma 2.1.1.

We now show part (e). Since $\overline{\text{span}}\{X_t, t \in T\} = H(X)$, then through the relationship $EX_tY = f(t)$, $t \in T$, the random variable Y is uniquely determined by f. Indeed, suppose $EX_tY_1 = f(t) = EX_tY_2$ for $Y_1, Y_2 \in H(X)$, then $Y_1 - Y_2$ is orthogonal to X_t for all $t \in T$, giving $Y_1 - Y_2 = 0$. Define $f(t) = EX_tY_f$ then $\{f : f(t) = EX_tY_f, Y_f \in H(X)\}$ is a Hilbert space of functions satisfying condition (RKHS2) of (1.6), if $\langle f, g \rangle_{K(C_X)} = \langle Y_f, Y_g \rangle_{L^2(\Omega,\mathcal{F},P)}$. Condition (RKHS1) of (1.6) obviously holds true. The fact that V is an isometry now follows.

To show (f) choose $\{\xi_i, i \in I\}$ an orthonormal basis in $H(X)$, then $\{\xi_i, i \in I\}$ are independent random variables by Proposition 2.2.1. For each $t \in T$, expand

$$X_t = \sum_{i \in I} \langle X_t, \xi_i \rangle_{L^2(\Omega,\mathcal{F},P)} \xi_i.$$

Then $e_i(t) = \langle X_t, \xi_i \rangle_{L^2(\Omega,\mathcal{F},P)}$ and $e_i = V^{-1}\xi_i$. Since $\sum_{i \in I} e_i^2(t) = C(t,t) < \infty$ we get the required convergences. $\qquad\square$

Lemma 2.2.2. *Let C be covariance on T. Then there exists a centered Gaussian random field $\{X_t, t \in T\}$ defined on a suitable probability space $(\Omega, \mathcal{F}, P)$ such that $C = C_X$. Moreover, we have the following Karhunen representation*

$$X_t = \sum_{i \in I} e_i(t) \xi_i, \tag{2.7}$$

where $\{e_i, i \in I\}$ is an orthonormal basis of the RKHS $K(C)$ and $\xi_i = \pi(e_i(\cdot))$. Here π is an isomorphism on $K(C)$ into a Gaussian space $H(X) \subseteq L^2(\Omega, \mathcal{F}, P)$ given by

$$X_t = \pi(C_t(\cdot)). \tag{2.8}$$

Proof. Let $K(C)$ be the RKHS of C. Then by Proposition 2.2.1, part (d), there exists an isomorphism π on $K(C)$ into a Gaussian space $H(X) \subseteq L^2(\Omega, \mathcal{F}, P)$. Define $X_t = \pi(C_t(\cdot))$ for each $t \in T$, then the set $\{X_t, t \in T\} \subseteq H(X)$ and hence it is a Gaussian random field. Clearly, for $t, s \in T$

$$C_X(t,s) = \langle X_t, X_s \rangle_{L^2(\Omega,\mathcal{F},P)} = \langle C_t(\cdot), C_s(\cdot) \rangle_{H(C)} = C(t,s).$$

Since

$$C_t(\cdot) = \sum_{i \in I} \langle C_t(\cdot), e_i \rangle_{K(C)} e_i(\cdot)$$

where $\{e_i, i \in I\}$ is an orthonormal basis of $K(C)$, we obtain $\langle C_t(\cdot), e_i \rangle_{K(C)} = e_i(t)$. Now with $\pi(e_i(\cdot)) = \xi_i$, we obtain the Karhunen expansion (2.7). $\qquad\square$

We now present some examples of Gaussian random fields.

Example 2.2.1.

(a) *Let* $T = \{(t_1, \cdots, t_n) = t,\ t_i \geq 0\}$ *and for* $t, t' \in T$,

$$C(t,t') = \prod_{i=1}^{n}(t_i \wedge t_i').$$

We can associate the covariance C with a centered Gaussian random field $\{X_t, t \in T\}$ using Lemma 2.2.2. This field is called Wiener-Lévy process or standard Brownian motion for $n = 1$, and for $n > 1$, Cameron–Yeh field.

(b) *Let* $T = \mathbb{R}^n$, *and for* $t, t' \in T$ *let*

$$C(t,t') = \frac{1}{2}\left(\|t\|_{\mathbb{R}^n}^{\alpha} + \|t'\|_{\mathbb{R}^n}^{\alpha} - \|t-t'\|_{\mathbb{R}^n}^{\alpha}\right), \quad 0 < \alpha \leq 2.$$

Using Exercise 1.1.4 part (c) and Exercise 1.1.3 part (a) we can see that C is a covariance. The associated centered Gaussian random field is called Lévy field for $\alpha = 1$, and for $\alpha \neq 1$, it is called fractional Brownian motion (fBm) for $n = 1$ and fractional Brownian field for $n > 1$ with the Hurst parameter $H = \alpha/2$.

(c) *For $T = S_n$ and $T = L_n$ (see Exercise 1.1.1) and $C(t,t') = \frac{1}{2}\{\psi(0,t') + \psi(0,t) - \psi(t,t')\}$, then the associated Gaussian field is called the Brownian motion on a sphere and on a hyperbolic plane, respectively.*

(d) *Let $T = C_0^{\infty}(\mathbb{R}^n)$ and $C(t,t') = \int_{\mathbb{R}^n} tt'\,d\lambda,\ t,t' \in T$. Then the associated (generalized) random field is called the white noise field of order 1.*

(e) *Let $T = C_0^{\infty}(\mathbb{R}^n)$ with $C(t,t') = \sum_{|\alpha| \leq p} \int_{\mathbb{R}^n} D^{\alpha}t D^{\alpha}t'\,d\lambda$. Then the associated Gaussian field is called the Sobolev white noise of order p.*

(f) *Let (S, Σ, μ) be a measurable space with a non-negative σ-finite measure μ, $T = \{A \in \Sigma, \mu(A) < \infty\}$, and $C(A,A') = \mu(A \cap A')$, $A, A' \in T$. Then C is a covariance on T. Then the associated Gaussian field $X(A)$, $A \in T$ is called an orthogonally (independently) scattered Gaussian set function. In this case, μ is called the variance measure of X.*

(g) *Let (S, Σ, μ) be a measurable space and $\{X(A), A \in \Sigma\}$ be a Gaussian field in $L^2(\Omega, \mathcal{F}, P)$, such that X is a countably additive $L^2(\Omega, \mathcal{F}, P)$–valued measure and $\mu(A \times B) = E(X(A)X(B))$ with μ of bounded variation on $\Sigma \times \Sigma$. Then we call X the Cramér field.*

Exercise 2.2.2. *Using Lemma 1.2.2 show that if C_1 and C_2 are two covariances on T with $C_1 \ll C_2$ and $\{X_i(t), t \in T\}$, $i = 1,2$ are associated Gaussian random fields, then*

$$X_1(t) = J \circ P(X_2(t))$$

where P is the projection of $H(X_2)$ onto a subspace M of $H(X_2)$ isomorphic to $H(X_1)$ with $J : M \to H(X_1)$ denoting the isomorphism.

Definition 2.2.3. *Let $\{X_t, t \in T\}$ be a Gaussian random field with covariance*

C_X. Then we call the isometry from $\pi : K(C_X) \to H(X)$ a stochastic integral. For $f \in K(C_X)$, $\pi(f)$ denotes the stochastic integral of f with respect to X.

Theorem 2.2.1. Let C be a covariance on a set T such that $L^2(S,\Sigma,\mu)$ is isometric to $H(C)$ for some σ-finite measure μ under an isometry V. Then $B(A) = \pi(V(1_A(\cdot)))$ for $A \in \Sigma$ such that $\mu(A) < \infty$ is a Gaussian set function and for each simple function of the form $f = \sum_{i=1}^{n} a_i 1_{A_i}(\cdot)$ we obtain

$$\pi \circ V(f) = \sum_{i=1}^{n} a_i B(A_i).$$

In this case for $f \in L^2(S,\Sigma,\mu)$ we denote

$$(\pi \circ V)(f) = \int_S f(u)\, dB(u). \tag{2.9}$$

Exercise 2.2.3. Prove Theorem 2.2.1.

2.3 Stochastic Integral Representation

Theorem 2.3.1. Let (S,Σ,μ) be measurable space with a finite measure μ and $\{f(t,\cdot),\ t \in T\}$ be a family of real-valued (complex-valued) functions generating the space $L^2(S,\Sigma,\mu)$, the real Hilbert space of real-valued (complex-valued) square integrable functions with respect to μ, with the inner product

$$\langle f,g \rangle_{L^2(S,\Sigma,\mu)} = Re\left(\int_S f\bar{g}\, d\mu \right).$$

If $C(t,t') = Re\left(\int_S \overline{f(t,u)} f(t',u)\, \mu(du) \right)$ then

$$K(C) = \left\{ h \,\middle|\, h(t) = Re\left(\int_S f(t,u)\bar{g}(u)\, \mu(du) \right),\ g \in L^2(S,\Sigma,\mu) \right\}.$$

The Gaussian random field associated with the covariance C is given by $X_t = \int f(t,u)\, dB(u)$, where $B(A) = \pi \circ V(1_A)$, and the isometry $V : L^2(S,\Sigma,\mu) \to K(C)$ is given by $V(g) = h$.

Proof. The first part follows from the property (RKHS2) in condition (1.6) and the second follows from Theorem 2.2.1 with the isometry $V(f(t,\cdot)) = C_t(\cdot)$. $\square$

Note that for the Wiener-Lévy process, by Example 1.2.1, we have

$$V : L^2(\mathbb{R}^n, \mathcal{B}(\mathbb{R}^n), \lambda) \to K(C), \quad (V(g))(s) = \int_0^{s_1} \dots \int_0^{s_n} g(u)\, du_1 \dots du_n.$$

Now using (2.8), we can define

$$B(t) = B([0,t]) = \pi\left(V(1_{[0,t]}(u))(\cdot)\right) = \pi(t \wedge \cdot) = \pi(C_t(\cdot)).$$

Hence,

$$B(t) = X_t, \quad \text{and} \quad X_t = \int_0^\infty 1_{[0,t]}(u)\, dB(u).$$

We present the following two important exercises regarding the Wiener–Lévy process.

Exercise 2.3.1. *Verify that*

(a) $X_0 = 0$

(b) $X_t - X_s$ *is independent of* $X_{t'} - X_{s'}$ *if* $(s,t) \cap [s',t'] = \varnothing$

(c) $X_t - X_s$ *and* $X_{t+h} - X_{s+h}$ *have the same distribution*

Exercise 2.3.2. *Using the representation in Theorem 2.3.1 and part (b) of Example 2.2.1, show that*

$$X_t = Re\left(k_n^{-1/2} \int_{\mathbb{R}^n} \frac{e^{itu} - 1}{\|u\|^{(n+1)/2}}\, dB(u) \right),$$

where B is a Gaussian set function with the Lebesgue measure as the variance measure.

We now leave it to the reader to prove that certain stochastic integrals defined in the literature are special cases of the stochastic integral defined above.

Exercise 2.3.3. *Let* $t_0 \in T$ *be such that* $C(t_0,t_0) = 0$, *then*

$$K(C) = \overline{\text{span}}\{C(\cdot,t) - C(\cdot,s), \quad s,t \in T\}.$$

Hint: To show the inclusion "⊆", use the fact that $C(\cdot,t) = C(\cdot,t) - C(\cdot,t_0)$.

Exercise 2.3.4.

(a) *This integral was defined in [45]. Let T be a subinterval of R and for* $I = [a,b], (\infty < a < b < \infty)$ *define*

$$C_X(\cdot,I) = C_X(\cdot,b) - C_X(\cdot,a), \quad I \subset T,$$

where C_X *is a covariance associated with a random field* $\{X_t, t \in T\}$. *Let*

$$M = \left\{ s : s = \sum_{k=1}^n u_k C(\cdot,I_k),\ \{u_1,\dots,u_n\} \subseteq \mathbb{R} \text{ and } I_1,\dots,I_n \text{ are subintervals of } T \right\}$$

Define $\Lambda_2(C_X)$ *as the completion of M in* $\|\cdot\|_{K(C_X)}$, *and* $\|\cdot\|_{\Lambda_2(C_X)}$ *as an extension of* $\|\cdot\|_{K(C_X)}$ *to* $\Lambda_2(C_X)$. *If* $X_{t_0} = 0$ *for some* $t_0 \in T$, *then* $\Lambda_2(C_X)$ *is isomorphic to* $K(C_X)$ *under the inner product* $\langle s,s'\rangle_{\Lambda_2(C_X)} = \langle s,s'\rangle_{K(C_X)}$. *Denote this isomorphism by V. Then show that for* $h \in \Lambda_2(C_X)$, *the stochastic integral S in [45] can be defined as* $S(h) = \pi(V(h))$.

(b) *Consider a Gaussian field* $\{X_A\}_{A \in \mathcal{M}}$ *with covariance* C_X, *where* $(M,\mathcal{M})$ *is a measurable space. Assume that the covariance* C_X *generates a signed measure of bounded variation on* $(M \times M, \mathcal{M} \otimes \mathcal{M})$, *through*

$$C(A \times B) = C_X(A,B), \quad A,B \in \mathcal{M}.$$

Sufficient conditions for a Gaussian process $\{X_t,\ t \in [0,T]\}$ to have such a covariance measure are given in [16] and will be discussed later. Define a marginal measure on $(M,\mathcal{M})$ by

$$\mu(A) = |C|(A \times M),$$

where $|C|$ is the total variation of C,

$$|C|(F) = \sup\left\{ \sum_{j=1}^{\infty} \mu(F_j) \ \middle|\ F_1, F_2, \ldots \in \mathcal{F} \text{ are disjoint}, \bigcup_{j=1}^{\infty} F_j \subseteq \mathcal{M} \right\}.$$

Since for $a_1, \ldots, a_j \in \mathbb{R}$ and $A_1, \ldots, A_n \in \mathcal{M}$

$$\sum_{i,j=1}^{n} a_i a_j C_X\left(A_i, A_j\right) = \sum_{i,j=1}^{n} a_i a_j C\left(A_i \times A_j\right)$$

$$= \int_{M \times M} \sum_{i,j=1}^{n} a_i a_j 1_{A_i}(u) 1_{A_j}(v) C(du, dv)$$

$$\leq \int_{M \times M} \sum_{i,j=1}^{n} \left| a_i a_j 1_{A_i}(u) 1_{A_j}(v) \right| |C|(du, dv)$$

$$= \int_{M \times M} \left| \sum_{i=1}^{n} a_i 1_{A_i}(u) \right| \left| \sum_{j=1}^{n} a_j 1_{A_j}(v) \right| |C|(du, dv)$$

$$\leq \left(\int_M \left(\sum_{i=1}^{n} a_i 1_{A_i}(u) \right)^2 |C|(du, dv) \right)^{\frac{1}{2}} \left(\int_M \left(\sum_{j=1}^{n} a_j 1_{A_j}(v) \right)^2 |C|(du, dv) \right)^{\frac{1}{2}}$$

$$= \int_M \left(\sum_{i=1}^{n} a_i 1_{A_i}(u) \right)^2 \mu(du)$$

$$= \int_M \left(\sum_{i,j=1}^{n} a_i a_j 1_{A_i}(u) 1_{A_j}(u) \right) \mu(du)$$

$$= \sum_{i,j=1}^{n} a_i a_j \mu(A_i \cap A_j)$$

Verify that $\tilde{C}_X(A,B) = \mu(A \cap B)$, $A, B \in \mathcal{M}$, is a covariance. The above calculations imply that $C_X \ll \tilde{C}_X$. Let us denote the Gaussian field associated with $\tilde{C}_X$ by $\tilde{X}(A)$, $A \in \mathcal{M}$. By Exercise 1.2.5 and Lemma 1.2.2, $K(C_X) \subseteq K(\tilde{C}_X)$ and there exists a Hilbert space $\tilde{H}$ and mappings $f_{\tilde{X}} : \mathcal{M} \to \tilde{H}$ and $f_X : \mathcal{M} \to \tilde{H}$, such that

$$\begin{aligned}
\langle f_{\tilde{X}}(A), f_{\tilde{X}}(B) \rangle_{\tilde{H}} &= \tilde{C}_X(A,B) \\
\langle f_X(A), f_X(B) \rangle_{\tilde{H}} &= C_X(A,B)
\end{aligned}$$

and in addition $f_X(A) = P(f_{\tilde{X}}(A))$, $A \in \mathcal{M}$, where P is an orthogonal projection of $\tilde{H}$ on $\overline{\text{span}}\{f_X(A), A \in \mathcal{M}\}$. Now

$$\tilde{\pi}(\tilde{C}_X(\cdot,A)) = \tilde{X}(A)$$
$$\pi(C_X(\cdot,A)) = X(A)$$

Let $\tilde{V}: \tilde{H} \to K(\tilde{C}_X)$ be defined by

$$\tilde{V}(f_{\tilde{X}}(A)) = \tilde{C}_X(\cdot,A).$$

Show that the mapping $\tilde{P}$ defined on $H(\tilde{X})$ by

$$\tilde{P}(\tilde{X}(A)) = \tilde{\pi}\tilde{V}Pf_{\tilde{X}}$$

is an orthogonal projection on a subspace of $H(\tilde{X})$ isomorphic to $\overline{\text{span}}\{X(A), A \in \mathcal{M}\}$, with the isometry $U\tilde{P}\tilde{X}(A) = X(A)$. For $f \in L^2(M,\mathcal{M},\mu)$ of the form $\sum_{i=1}^{n} a_i 1_{A_k}$, define

$$\tilde{S}(f) = \sum_{i=1}^{n} a_i \tilde{X}(A_k).$$

If $S(f) = \sum_{i=1}^{n} a_i X(A_k)$, then $S(f) = U\tilde{P}\tilde{S}(f)$, and for a Cauchy sequence $\{f_n\}_{n=1}^{\infty} \subseteq L^2(M,\mathcal{M},\mu)$

$$E\left(S(f_n) - S(f_m)\right)^2 \le E\left(\tilde{S}(f_n) - \tilde{S}(f_m)\right)^2.$$

Therefore, the integral S can be defined for all functions $f \in L^2(M,\mathcal{M},\mu)$. In fact, the integral S can be extended to the completion of $L^2(M,\mathcal{M},\mu)$ in the norm $(ES(\cdot))^2)^{1/2}$.

Remark 2.3.1. *The random field $\{\tilde{X}(A), A \in \mathcal{M}\}$ defined in Exercise 2.3.4 is an orthogonally scattered Gaussian random measure, that is,*

$$E\left(\tilde{X}(A)\tilde{X}(B)\right) = 0 \quad \text{if } A \cap B = \varnothing.$$

The construction in Exercise 2.3.4 defines a stochastic integral S with respect to a projection X of an orthogonally scattered Gaussian random measure $\tilde{X}$.

2.4 Chaos Expansion

Let $\{X_t\}_{t \in T}$ be a centered Gaussian field on a probability space $(\Omega, \mathcal{F}, P)$ and consider $H(X) = \overline{\text{span}}\{X_t, t \in T\}$. Denote σ-field $\sigma(H(X))$ by $\mathcal{F}_X$. Let for $Z \in H(X)$

$$d_Z = e^{Z - \frac{1}{2}EZ^2}.$$

We shall show that $d_Z \in L^2(\Omega, \mathcal{F}_X, P)$ and that the family $\{d_Z, z \in H(X)\}$ is dense in $L^2(\Omega, \mathcal{F}_X, P)$. We denote by $\{\xi_i, i \in I\}$ an orthonormal basis in $H(X)$. To show that d_Z is square integrable we calculate

$$E e^{2Z - EZ^2} = e^{EZ^2} E e^{2z} = e^{EZ^2} e^{(4)(\frac{1}{2}EZ^2)}.$$

Clearly $\overline{\text{span}}\{d_Z, Z \in H(X)\} \subseteq L^2(\Omega, \mathcal{F}_X, P)$. Suppose there exists an $f \in L^2(\Omega, \mathcal{F}_X, P)$ orthogonal to all d_Z, that is,

$$\int_\Omega e^{Z - \frac{1}{2}EZ^2} f \, dP = 0 \quad \text{for all} \quad Z \in H(X).$$

In particular, for any finite subset $N \subseteq I$, we have

$$\int_\Omega e^{\Sigma_{i \in N} a_i \xi_i - \frac{1}{2} \Sigma_{i \in N} a_i^2} f \, dP = 0. \tag{2.10}$$

If we denote $\mathcal{F}_N = \sigma\{\xi_i, i \in N\}$, then we obtain from (2.10) that

$$\int_\Omega e^{\Sigma_{i \in N} a_i \xi_i} E^{\mathcal{F}_N} f \, dP = 0.$$

By the finite dimensional result $E^{\mathcal{F}_N} f = 0$, P–a.e. Hence, by the Martingale Convergence Theorem $f = 0$, P–a.e.

Recall $\{h_n(x), n = 0, 1, ...\}$, the Hermite polynomials defined in (2.5), and denote

$$H_n(x) = (\sqrt{n!}) h_n(x). \tag{2.11}$$

Exercise 2.4.1. *Show that the Hermite polynomials $H_n(x)$ are given by the exponential generating function*

$$e^{ux - \frac{1}{2}u^2} = \sum_{n=0}^{\infty} \frac{u^n}{n!} H_n(x).$$

Prove that

$$H_n(x) = (-1)^n e^{\frac{x^2}{2}} \frac{d^n}{dx^n} e^{-\frac{x^2}{2}}.$$

By Exercise 2.4.1, for a set N of finite cardinality,

$$\prod_{i \in N} e^{a_i \xi_i - \frac{1}{2} a_i^2} = \prod_{i \in N} \sum_{n_i=0}^{\infty} \frac{a_i^{n_i}}{n_i!} H_{n_i}(\xi_i). \tag{2.12}$$

Let us now assume that I is countable, or equivalently, that $H(X)$ is separable. Then it follows from (2.12) that the system

$$\left\{ \prod_{i \in I} \frac{1}{\sqrt{n_i!}} H_{n_i}(\xi_i), \sum_{i \in I} n_i = n, n = 0, 1, ... \right\} \tag{2.13}$$

generates $L^2(\Omega, \mathcal{F}_X, P)$.

Exercise 2.4.2. (a) *Prove that the system defined in (2.13) is orthonormal in* $L^2(\Omega, \mathcal{F}_X, P)$.

(b) *Show that the systems*

$$\left\{ \prod_{i \in I} \frac{1}{\sqrt{n_i!}} H_{n_i}(\xi_i), \sum_{i \in I} n_i = n \right\} \quad and \quad \left\{ \prod_{i \in I} \frac{1}{\sqrt{m_i!}} H_{m_i}(\xi_i), \sum_{i \in I} m_i = m \right\}$$

are mutually orthogonal for $n \neq m$.

Hint: Note that if Z_n *and* Z_m *are any two elements of the first and the second system, respectively, then the inequality* $n \neq m$ *implies that there exists* $i_0 \in I$, *such that* $n_{i_0} \neq m_{i_0}$.

Denote $H^n = \overline{\text{span}} \{ \prod_{i \in I} H_{n_i}(\xi_i), \sum_{i \in I} n_i = n \}$. We are ready to present the chaos expansion due to Wiener [122].

Theorem 2.4.1. *Let* $H = H(X)$ *be a separable Gaussian space. Then*

$$L^2(\Omega, \mathcal{F}_X, P) = \bigoplus_{n=0}^{\infty} H^n. \tag{2.14}$$

In particular, each function $f \in L^2(\Omega, \mathcal{F}_X, P)$ *has the following expansion*

$$f = Ef + \sum_{n=1}^{\infty} \sum_{\substack{i_1, \dots, i_r \\ n_{i_1}, \dots, n_{i_r}}} a_{n_{i_1} \dots n_{i_r}}^{i_1 \dots i_r} \prod_{k=1}^{n} H_{n_{i_k}}(\xi_{i_k}), \tag{2.15}$$

where the inner sum is over all finite subsets $(i_1, \dots, i_r)$ *of* I *and* r-tuples of positive integers $(n_{i_1}, \dots, n_{i_r})$, *such that* $\sum_{k=1}^{r} n_{i_k} = n$, $n \geq 1$.

The series (2.15) converges in $L^2(\Omega, \mathcal{F}_X, P)$.

Exercise 2.4.3. *Show that* $H^n = H^{\otimes n}(X)$, *that is,* $H(X)$ *is the* n-fold tensor product of $H(X)$.

Chapter 3

Stochastic Integration for Gaussian Random Fields

3.1 Multiple Stochastic Integrals

In Chapter 1, we introduced the concept of the tensor product

$$K(C_1) \otimes ... \otimes K(C_p) = K(C_1 \otimes ... \otimes C_p)$$

of RKHSs $K(C_i)$ and covariances $C_1, ..., C_p$, defined, respectively, on sets $T_1, ..., T_p$. We recall that $K(C_1) \otimes ... \otimes K(C_p)$ consists of functions f of the form

$$f(t_1, ..., t_p) = \sum_{\alpha_1, ..., \alpha_p} a_{\alpha_1, ..., \alpha_p} e^1_{\alpha_1}(t_1) ... e^p_{\alpha_p}(t_p), \qquad (3.1)$$

such that $\sum_{\alpha_1, ..., \alpha_p} a^2_{\alpha_1, ..., \alpha_p}$ is finite and $\left\{ e^j_{\alpha_j}, \alpha_j \in J \right\}$ is an orthonormal basis in $K(C_j)$, $j = 1, ..., p$. The scalar product of two such functions is defined by

$$\langle f, g \rangle_{K(C_1 \otimes ... \otimes C_p)} = \sum_{\alpha_1, ..., \alpha_p} a_{\alpha_1, ..., \alpha_p} b_{\alpha_1, ..., \alpha_p}$$

where the function g has the form (3.1) with the coefficients $b_{\alpha_1, ..., \alpha_p}$ replacing $a_{\alpha_1, ..., \alpha_p}$. We have the following lemma.

Lemma 3.1.1. (a) *Let $f \in K(C_1 \otimes ... \otimes C_p)$ and $g \in K(C_{p+1} \otimes ... \otimes C_{p+q})$, then the function*

$$(fg)(t_1, ..., t_{p+q}) = f(t_1, ..., t_p) g(t_{p+1}, ..., t_{p+q})$$

belongs to $K(C_1 \otimes ... \otimes C_{p+q})$.

(b) *In the case when $C_1 = C_2 = ... = C_p = C$, denoting $K(C)^{\otimes p} = K(C \otimes, ..., \otimes C)$, if $f \in K(C)^{\otimes p}$, then for any permutation π of the set $\{1, ..., p\}$, the function*

$$f^\pi(t_1, ..., t_p) = f(t_{\pi(1)}, ..., t_{\pi(p)})$$

belongs to $K(C)^{\otimes p}$.

Exercise 3.1.1. *Prove Lemma 3.1.1.*

We call $K(C)^{\otimes p}$ the pth tensor product of $K(C)$, and denote it by $K^{\otimes p}$ if there is no confusion about the covariance. For a function $f \in K^{\otimes p}$, we define and denote its symmetrization by

$$\tilde{f}(t_1,...,t_p) = \frac{1}{p!}\sum_{\pi} f(t_{\pi(1)},...,t_{\pi(p)})$$

where the sum is over all permutations of the set $\{1,...,p\}$. We note that $\tilde{f} \in K^{\otimes p}$ and is symmetric. The subspace of $K^{\otimes p}$ consisting of all symmetric functions will be denoted by $K^{\odot p}$. Clearly $\tilde{f} \in K^{\odot p}$ for all $f \in K^{\otimes p}$. For $p = 0$, $K^{\otimes 0} = K^{\odot 0} = \mathbb{R}$.

The stochastic integral $\pi : K \to L^2(\Omega,\mathcal{F},P)$ in Definition 2.2.3 is an isometry and for $f \in K$, we denote it by

$$\pi(f) = I_1(f),$$

and call it the order one stochastic multiple integral.

We now define stochastic multiple integrals I_p of order p, for any integer p. For $p = 0$,

$$I_0(f) = f, \ f \in K^{\otimes p0} = \mathbb{R}.$$

Note that

$$I_0 : K^{\otimes 0} \ \to \ L^2(\Omega,\mathcal{F},P),$$
$$I_1 : K^{\otimes 1} \ \to \ L^2(\Omega,\mathcal{F},P).$$

We define the order $p + 1$ stochastic multiple integral I_{p+1} as a map from $K^{\otimes p+1} \to L^2(\Omega,\mathcal{F},P)$. Because of the form of the elements of $K^{\otimes p+1}$, it suffices to define I_{p+1} for functions of the form

$$f(t_1,...,t_p,t_{p+1}) = e_{\alpha_1}(t_1) \cdot e_{\alpha_p}(t_p)e_{\alpha_{p+1}}(t_{p+1}), \tag{3.2}$$

where among the indexes $\alpha_1,...\alpha_{p+1}$, n are different, $n = 1,...,p+1$, with repeats $p_1,...,p_n$, so that $p_1 + ... + p_n = p + 1$. Let us denote the corresponding n different elements e_{α_i} by $u_1,...,u_n \in K$. Define for f as in (3.2)

$$I_{p+1}(f) = \prod_{i=1}^{n} H_{p_i}(I_1(u_i)), \tag{3.3}$$

where H_{p_i} are Hermite polynomials of order p_i defined in (2.11).

We extend the operator I_{p+1} linearly to the linear manifold generated by elements of the form (3.2). We need the following lemma to extend I_{p+1} to the entire space $K^{\otimes p+1}$.

Lemma 3.1.2. *Let $f,g \in K^{\otimes p+1}$ be of the form (3.2). Then for $p \geq 0$,*
(a) $I_{p+1}(\tilde{f}) = I_{p+1}(f)$.

(b) $E\left(I_{p+1}(f)\right) = 0$ *and*

$$E\left(I_{p+1}(f)I_{p+1}(g)\right) = (p+1)!\langle \tilde{f},\tilde{g}\rangle_{K^{\otimes p+1}} \tag{3.4}$$

$$= \begin{cases} (p+1)!\|\tilde{f}\|^2_{K^{\otimes p+1}} & \text{if } f = g \\ 0 & \text{if } f \neq g \end{cases}$$

Proof. Part (a) follows by simple observation that each component of the sum defining $\tilde{f}$ has the same image. To prove (b) note that if $f \neq g$, then $I_{p+1}(f)$ and $I_{p+1}(g)$ involve two different order Hermite polynomials or two first order integrals of different basis elements of $K^{\otimes p+1}$ giving that the LHS of (3.4) is zero. Also, the the scalar product $\langle \tilde{f},\tilde{g}\rangle_{K^{\otimes p+1}}$ is zero since at least one p_i corresponding to e_{α_i} does not match. The fact that $E\left(I_{p+1}(f)\right) = 0$ follows, since $H_0\left(I_1(u_i)\right) = 1$.

Assume now that

$$f = g = u_1(t_1)...u_1\left(t_{p_1}\right)u_2\left(t_{p_1+1}\right)...u_2\left(t_{p_1+p_2}\right)...u_n\left(t_{p_1+...+p_n}\right),$$

where u_n are distinct basis elements of $K^{\otimes p+1}$. Then

$$\|\tilde{f}\|^2_{K^{\otimes p+1}} = \langle \tilde{f},\tilde{f}\rangle_{K^{\otimes p+1}} = \left(\frac{1}{(p+1)!}\right)^2 \sum_{\pi,\pi'}\langle f^\pi, f^{\pi'}\rangle_{K^{\otimes p+1}}$$

$$= \left(\frac{1}{(p+1)!}\right)^2 \sum_{\pi,\pi'}(1),$$

where the summation is over all $(p+1)!$ choices of the permutation π and $p_1!...p_n!$ choices of permutation π' resulting in a unit scalar product. Thus,

$$(p+1)!\|\tilde{f}\|^2_{K^{\otimes p+1}} = p_1!...p_n! = E\left(\prod_{i=1}^n H_{p_i}\left(I_1(u_i)\right)\right)^2 = E\left(I_{p+1}(f)\right)^2.$$

$\square$

We conclude from Lemma 3.1.2 that for basis elements $\tilde{\varphi}$, $\tilde{\psi} \in K^{\otimes p+1}$, the map

$$U_{p+1} = ((p+1)!)^{-\frac{1}{2}} I_{p+1} : K^{\otimes p+1} \rightarrow L^2(\Omega,\mathcal{F},P)$$

is unitary. Hence, it extends to a unitary map

$$\tilde{U}_{p+1} : K^{\odot p+1} \rightarrow L^2(\Omega,\mathcal{F},P).$$

Definition 3.1.1. *For any non-negative integer p, we define*

$$I_p = (p!)^{\frac{1}{2}}\tilde{U}_p : K^{\odot p+1} \rightarrow L^2(\Omega,\mathcal{F},P)$$

and call this map the multiple Wiener integral of order p with respect to the centered Gaussian random field X_t, $t \in T$. Clearly $I_p(f) = I_p(\tilde{f})$ defines the multiple Wiener integral on $K^{\otimes p}$.

Theorem 3.1.1. *Let I_p, $p \geq 0$ be the multiple Wiener integral in Defini-tion 3.1.1. Then for $f, g \in K^{\otimes p}$,*

(a) *For $p > 0$, $E\left(I_p(f)\right) = 0$ and*

$$E\left(I_p(f)I_p(g)\right) = p!\langle \tilde{f}, \tilde{g} \rangle_{K^{\otimes p}} = \begin{cases} p!\|\tilde{f}\|^2_{K^{\otimes p}} & \text{if } f = g \\ 0 & \text{if } f \neq g \end{cases} \tag{3.5}$$

(b) *$I_p\left(K^{\otimes p}\right) \perp I_k\left(K^{\otimes p}\right)$ for $p \neq p$.*

Proof. By Lemma 3.1.2 and the orthogonality of $H_p(I(u_i))$ and $H_0\left(I_1(u_i)\right) = 1$, the theorem is true for functions of the form (3.2). For a general function f, $I_p(f)$ is a limit of $I_p(f_n)$ in $L^2(\Omega, \mathcal{F}, P)$ and hence in $L^1(\Omega, \mathcal{F}, P)$ with f_n a finite linear combination of elements of the form 3.2, and hence the first claim in part (a) follows. The second claim in (a) follows from the fact that $(p!)^{-1/2}I_p$ is an isometry.

Part (b) follows from the fact that $I_p(f_p) \perp I_k(f_k)$ for elements f_p of the basis of $K^{\otimes p}$ and elements f_k of the basis of $K^{\otimes k}$ for $p \neq k$. $\qquad\square$

Let us now consider f of the form (3.2). Denote $\omega_i = I_1(u_i)$ and assume that $e_{\alpha_{p+1}}(t) = u_j(t)$ for some j. Then using the property of Hermite polynomials

$$H_n(t) = tH_{n-1}(t) - (n-1)H_{n-2}(t)$$

we can calculate

$$
\begin{aligned}
I_{p+1}(f) &= \prod_{i=1}^{n} H_{p_i}(\omega_i) \tag{3.6} \\
&= H_{p_1}(\omega_1)...H_{p_{j-1}}(\omega_{j-1})\left(\omega_j H_{p_j-1}(\omega_j) - (p_j-1)H_{p_j-2}(\omega_j)\right) \\
&\quad H_{p_{j+1}}(\omega_{j+1})...H_{p_n}(\omega_n) \\
&= I_p(g)I_1(h) - \sum_{k=1}^{p} I_{p-1}\left(g \underset{k}{\otimes} h\right),
\end{aligned}
$$

where $g(t_1,...,t_p) = e_{\alpha_1}(t_1)...e_{\alpha_p}(t_p)$, $h(t) = e_{\alpha_{p+1}}(t)$ and

$$\left(g \underset{k}{\otimes} h\right)(t_1,...,t_{k-1},t_{k+1},...,t_p) - \langle g(t_1,...,t_{k-1},\cdot,t_{k+1},...,t_p), h(\cdot)\rangle_K.$$

Given $g \in K^{\otimes p}$, $h \in K$ of the form

$$
\begin{aligned}
g(t_1,...,t_p) &= \sum_{\alpha_1,...,\alpha_p} a_{\alpha_1,...,\alpha_p} e_{\alpha_1}(t_1)...e_{\alpha_p}(t_p) \\
h(t) &= \sum_{\beta} b_{\beta} e_{\beta}(t)
\end{aligned}
$$

we have

$$(gh)\big((t_1,...,t_{p+1})\big) = \sum_{\alpha_1,...,\alpha_p,\beta} a_{\alpha_1,...,\alpha_p} b_\beta e_{\alpha_1}(t_1)...e_{\alpha_p}(t_p)e_\beta(t_\beta).$$

Using the result of the calculation in (3.6) we obtain

$$I_{p+1}(gh) = I_p(g)I_1(h) - \sum_{k=1}^{p} I_{p-1}\left(g \underset{k}{\otimes} h\right). \tag{3.7}$$

We remark that it can be concluded from the property (3.7) that for f of the form (3.2), the multiple Wiener integral I_{p+1} satisfies Equation (3.3) and hence the multiple Wiener integral I_p defined in Definition 3.1.1 coincides with the integral defined in [48].

Recall that from the proof of Lemma 2.2.2, $\{\omega_\alpha = I_1(e_\alpha), \alpha \in J\}$ are i.i.d. standard normal random variables forming an ONB in $H(X)$ corresponding to $\{e_\alpha, \alpha \in J\}$, which is a complete orthonormal system in $K(C)$. According to (3.3) and Lemma 3.1.2, the linear space spanned by all random variables of type $\prod_{i=1}^{n} H_{p_i}(I_1(e_{\alpha_i}))$ is the same as the image $I_p\left(K^{\odot p}\right)$, where H_{p_i} are Hermite polynomials of order p_i defined in (2.11) and p_i are non-negative integers whose sum equals p. From Theorem 2.4.1 we have the following lemma.

Lemma 3.1.3. *Let $K = K(C)$ be a separable RKHS. Then*

$$L^2(\Omega,\mathcal{F}_X,P) = \sum_{p=0}^{\infty} \bigoplus I_p\left(K^{\odot p}\right). \tag{3.8}$$

3.2 Skorokhod Integral

Let $\{X_t, t \in T\}$ be a centered Gaussian field defined on a probability space $(\Omega,\mathcal{F},P)$ with covariance C, where $\mathcal{F} = \sigma\{X_t,; t \in T\}$. We now use Lemma 3.1.3 to define the Skorokhod integral of a real-valued random field $\{g_t(\omega), t \in T\}$. We assume that $g : (\Omega,\mathcal{F}) \to (K(C),\mathcal{B}(K(C)))$ is a *Bochner measurable* function; that is, $g(\omega) = \lim_{n\to\infty} g_n(\omega)$ in $K(C)$, where $g_n : \Omega \to K(C)$ are simple functions and $g_n^{-1}(k) \in \mathcal{F}, k \in K(C)$. Then by [18], there exists a set $N \in \mathcal{F}$ such that $P(N) = 0$ and for $\omega \in N^c$, $g.(\omega)$ takes values in a separable subspace of K. Hence, by choosing an orthonormal basis of this separable subspace, if necessary, we can assume that $K(C)$ is separable.

Consider $g(\cdot) \in L^2(\Omega,K(C))$. Since

$$|g_t(\omega)| = \left|\langle g.(\omega),C(\cdot,t)\rangle_{K(C)}\right| \le \|g.(\omega)\|^2_{K(C)} C(t,t)$$

we conclude that for each $t \in T$, $E|g_t(\omega)|^2 < \infty$. By Lemma 3.1.3, for each $t \in T$, there exists a unique sequence $\{f_p^t(\cdot)\}_{p=0}^{\infty} \subseteq K^{\odot p}$, such that

$$g_t(\omega) = \sum_{p=0}^{\infty} I_p\left(f_p^t\right)(\omega). \tag{3.9}$$

Lemma 3.2.1. *If $g_t(\omega)$ has the decomposition (3.9) then:*

(a) *Functions $f_p(t_1,...,t_p,t) = f_p^t(t_1,...,t_p)$ can be chosen in such a way that $f_p(t_1,...,t_p,t) \in K^{\otimes p+1}$ and are symmetric in the first p variables.*

(b) *The representation in (a) is unique, that is, if*

$$g_t(\omega) = \sum_{p=0}^{\infty} I_p\left(f_p(\cdot,t)\right)(\omega) = \sum_{p=0}^{\infty} I_p\left(f_p'(\cdot,t)\right)(\omega)$$

and f_p, f_p' satisfy the symmetry requirement in (a), then $f_p(t_1,...,t_p,t) = f_p'(t_1,...,t_p,t)$ for all $(t_1,...,t_p,t)$.

Proof. If $g_t^n(\omega) = \sum_{k=1}^n F_k(\omega)g_k(t)$ with $F_k(\cdot) \in L^2(\Omega)$ and $g_k \in K$, then

$$g_t^n = \sum_{k=1}^n \sum_{p=0}^{\infty} I_p\left(f_{k,p}\right)g_k(t) \quad = \quad \sum_{p=0}^{\infty}\sum_{k=1}^n I_p\left(f_{k,p}\right)g_k(t)$$

$$= \quad \sum_{p=0}^{\infty} I_p\left(\sum_{k=1}^n f_{k,p}g_k(t)\right)$$

In general, since $g : \Omega \to K$ is Bochner measurable, it can be approximated by K-valued random variables g_t^n as above, and we can write

$$g_t^n(\omega) = \sum_{p=0}^{\infty} I_p\left(f_{n,p}(\cdot,t)\right)(\omega),$$

where $f_{n,p}(\cdot,\cdot) \in K^{\otimes p+1}$ and for a fixed t, $f_{n,p}(\cdot,t) \in K^{\odot p}$. The approximation is in $L^2(\Omega,K)$, $E\,\|g - g^n\|_K^2 \to 0$ as $n \to \infty$. We will use the fact that g^n is a Cauchy sequence in $L^2(\Omega,K)$, $E\,\|g^m - g^n\|_K^2 \to 0$ as $m,n \to \infty$. Let

$$f_{n,p}\left(t_1,...,t_p,t\right) - f_{m,p}\left(t_1,...,t_p,t\right)$$

$$= \sum_{\alpha_1,...,\alpha_{p+1}} a_{\alpha_1,...,\alpha_{p+1}}^{n,m,p} e_{\alpha_1}(t_1)...e_{\alpha_p}(t_p)e_{\alpha_{p+1}}(t)$$

$$= \sum_{\alpha_{p+1}}\left(\sum_{\alpha_1,...,\alpha_p} a_{\alpha_1,...,\alpha_{p+1}}^{n,m,p} e_{\alpha_1}(t_1)...e_{\alpha_p}(t_p)\right)e_{\alpha_{p+1}}(t).$$

Then, with the notation indicating that the norm in K is taken of the function depending on t, we have

$$E\,\|g_t^m - g_t^n\|_K^2 = E\left\|\sum_{p=0}^{\infty} I_p\left(f_{n,p}(\cdot,t) - f_{m,p}(\cdot,t)\right)\right\|_K^2$$

$$= E\left(\sum_{p,p'=0}^{\infty} \langle I_p\left(f_{n,p}(\cdot,t) - f_{m,p}(\cdot,t)\right), I_{p'}\left(f_{n,p'}(\cdot,t) - f_{m,p'}(\cdot,t)\right)\rangle_K\right)$$

$$= \sum_{p=0}^{\infty} E \left\| I_p \left(f_{n,p}(\cdot,t) - f_{m,p}(\cdot,t) \right) \right\|_K^2.$$

But

$$E \left\| I_p \left(f_{n,p}(\cdot,t) - f_{m,p}(\cdot,t) \right) \right\|_K^2$$

$$= E \sum_{\alpha_{p+1}} \left\langle I_p \left(f_{n,p}(\cdot,t) - f_{m,p}(\cdot,t) \right), e_{\alpha_{p+1}}(t) \right\rangle_K^2$$

$$= \sum_{\alpha_{p+1}} E \left(\sum_{\alpha_1,\ldots,\alpha_p} a_{\alpha_1,\ldots,\alpha_{p+1}}^{n,m,p} I_p \left(e_{\alpha_1}(t_1)\ldots e_{\alpha_p}(t_p) \right) \right)^2$$

$$= \sum_{\alpha_{p+1}} p! \sum_{\alpha_1,\ldots,\alpha_p} \left(a_{\alpha_1,\ldots,\alpha_{p+1}}^{n,m,p} \right)^2$$

$$= p! \sum_{\alpha_1,\ldots,\alpha_{p+1}} \left(a_{\alpha_1,\ldots,\alpha_{p+1}}^{n,m,p} \right)^2$$

$$= p! \left\| f_{n,p}(\cdot,t) - f_{m,p}(\cdot,t) \right\|_{K^{\otimes p+1}}^2.$$

This implies that

$$E \left\| g_t^m - g_t^n \right\|_K^2 = \sum_{p=0}^{\infty} p! \left\| f_{n,p}(\cdot,t) - f_{m,p}(\cdot,t) \right\|_{K^{\otimes p+1}}^2 \to 0, \quad \text{as } m,n \to \infty. \quad (3.10)$$

Hence, for each p,

$$\left\| f_{n,p}(\cdot,t) - f_{m,p}(\cdot,t) \right\|_{K^{\otimes p+1}}^2 \to 0, \quad \text{as } m,n \to \infty,$$

so that $\left\{ f_{n,p} \right\}_{n=1}^{\infty} \subseteq K^{\otimes p+1}$ is a Cauchy sequence. Let $f_p = \lim_{n \to \infty} f_{n,p}$ in $K^{\otimes p+1}$. Define

$$h_t(\omega) = \sum_{p=0}^{\infty} I_p \left(f_p(\cdot,t) \right),$$

then, using similar calculations as above, we obtain that

$$E \left\| g_t^n - h_t \right\|_K^2$$

$$= E \left\| \sum_{p=0}^{\infty} I_p \left(f_{n,p}(\cdot,t) - f_p(\cdot,t) \right) \right\|_K^2$$

$$= \sum_{p=0}^{\infty} E \left\| I_p \left(f_{n,p}(\cdot,t) - f_p(\cdot,t) \right) \right\|_K^2$$

$$= \sum_{p=0}^{\infty} p! \left\| f_{n,p}(\cdot,t) - f_p(\cdot,t) \right\|_{K^{\otimes p+1}}^2 \to 0, \quad \text{as } n \to \infty,$$

where the convergence to zero is implied by (3.10) and is left for the reader to prove in Exercise 3.2.1. This proves that $g_t(\omega) = h_t(\omega)$. Functions f_p have the properties desired in (a).

To show part (b), note that

$$
\begin{aligned}
0 &= E \left\| \sum_{p=0}^{\infty} I_p\left(f_p(\cdot,t)\right) - \sum_{p=0}^{\infty} I_p\left(f_p'(\cdot,t)\right) \right\|_K^2 \\
&= \sum_{p=0}^{\infty} E \left\| I_p\left(f_p(\cdot,t) - f_p'(\cdot,t)\right) \right\|_K^2 \\
&= \sum_{p=0}^{\infty} p! \left\| f_p(\cdot,t) - f_p'(\cdot,t) \right\|_{K^{\otimes p+1}}^2
\end{aligned}
$$

$\square$

Exercise 3.2.1. *Show that in the proof of Lemma 3.2.1*

$$
\sum_{p=0}^{\infty} p! \left\| f_{n,p}(\cdot,t) - f_p(\cdot,t) \right\|_{K^{\otimes p+1}}^2 \to 0, \quad as \ n \to \infty.
$$

Remark 3.2.1. *As a consequence of the proof of Lemma 3.2.1, we have that if a field $g \in L^2(\Omega,K)$ has a decomposition*

$$
g_t(\omega) = \sum_{p=0}^{\infty} I_p\left(f_p(\cdot,t)\right),
$$

where $f_p(\cdot,\cdot) \in K^{\otimes p+1}$ and $f_p(\cdot,t) \in K^{\odot p}$ for each $t \in T$, then

$$
E\|g\|_K^2 = \sum_{p=0}^{\infty} p! \|f_p\|_{K^{\otimes p+1}}. \tag{3.11}
$$

In fact, we have the following lemma.

Lemma 3.2.2. *Let $f(\cdot,\cdot) \in K^{\otimes p+1}$, $g(\cdot,\cdot) \in K^{\otimes q+1}$ and for each $t \in T$, $f(\cdot,t) \in K^{\odot p}$ and $g(\cdot,t) \in K^{\odot q}$. Then for $u_t(\omega) = I_p(f(\cdot,t))$ and $v_t(\omega) = I_q(g(\cdot,t))$,*

$$
E\langle u,v \rangle_K = \begin{cases} p! \langle f,g \rangle_{K^{\otimes p+1}} & \text{if } p = q \\ 0 & \text{if } p \neq q \end{cases} \tag{3.12}
$$

Exercise 3.2.2. *Prove Lemma 3.2.2.*

We now define the Skorokhod integral.

Definition 3.2.1. *Let $\{X_t,\ t \in T\}$ be a centered Gaussian random field with covariance C and $K = K(C)$ be the associated RKHS. Assume that $u \in L^2(\Omega,K)$ has a decomposition*

$$
u_t(\omega) = \sum_{p=0}^{\infty} I_p\left(f_p(\cdot,t)\right),
$$

where $f_p(\cdot,\cdot) \in K^{\otimes p+1}$ and $f_p(\cdot,t) \in K^{\odot p}$ for each $t \in T$, and hence such f_p is unique due to Lemma 3.2.1. Denote by $\tilde{f}_p(\cdot)$ the symmetrization of f_p as an element of $K^{\otimes p+1}$. If the series

$$I^s(u) = \sum_{p=0}^{\infty} I_{p+1}\left(f_p\right) = \sum_{p=0}^{\infty} I_{p+1}\left(\tilde{f}_p\right) \tag{3.13}$$

converges in $L^2(\Omega)$, then we say that the function u is Skorokhod integrable and we call $I^s(u)$ the Skorokhod integral of u with respect to the Gaussian field X.

Lemma 3.2.3. *With the notation of Definition 3.2.1, the Skorokhod integral $I^s(u) \in L^2(\Omega)$ if and only if*

$$\sum_{p=0}^{\infty} (p+1)! \left\|\tilde{f}_p\right\|^2_{K^{\otimes p+1}} < \infty. \tag{3.14}$$

In this case, $\|I^s(u)\|^2_{L^2(\Omega)}$ coincides with the sum in (3.14).

Exercise 3.2.3. *Prove Lemma 3.2.3.*

In order to connect the Skorokhod integral with other anticipative integrals, we need the concept of differentiation due to Skorokhod. In Chapter 4, we study the relationship of this differentiation with the one introduced by Malliavin.

3.3 Skorokhod Differentiation

Let $\{X_t, t \in T\}$ be a centered Gaussian field defined on a probability space $(\Omega, \mathcal{F}, P)$ with covariance C, where $\mathcal{F} = \sigma\{X_t, ; t \in T\}$. Let us recall the spaces $L^2(\Omega)$ and $L^2(\Omega, K)$ of P-square integrable real-valued and K-valued Bochner measurable functions with respect to the σ-field $\mathcal{F}$. Similar as with the Skorokhod integral, using Bochner measurability, we can assume that the RKHS $K = K(C)$ is separable.

Definition 3.3.1. *Let $F \in L^2(\Omega)$ have the unique decomposition $F = \sum_{p=0}^{\infty} I_p\left(f_p\right)$, with $f_p \in K^{\odot p}$. Define*

$$D_t F = \sum_{p=1}^{\infty} p I_{p-1}\left(f_p(\cdot,t)\right) \tag{3.15}$$

if the series converges in $L^2(\Omega, K)$. We call $D_t F \in L^2(\Omega, K)$ the Skorokhod derivative of F.

Since

$$E\left\|\sum_{p=1}^{\infty} p I_{p-1}\left(f_p(\cdot,t)\right)\right\|^2_K = \sum_{p=1}^{\infty} p^2(p-1)! \left\|f_p\right\|^2_{K^{\otimes p}}$$

$$= \sum_{p=1}^{\infty} pp! \|f_p\|_{K^{\otimes p}}^2,$$

we have the following condition for the existence of the Skorokhod derivative.

Lemma 3.3.1. *Let* $F = \sum_{p=0}^{\infty} I_p(f_p) \in L^2(\Omega)$, *with* $f_p \in K^{\odot p}$. *Then* $D_t F \in L^2(\Omega, K)$ *if and only if*

$$\sum_{p=1}^{\infty} pp! \|f_p\|_{K^{\otimes p}}^2 < \infty. \qquad (3.16)$$

Definition 3.3.2. *We define the Skorokhod derivative* D *as an operator from* $L^2(\Omega)$ *to* $L^2(\Omega, K(C))$ *whose domain consists of functions satisfying condition* (3.16), *and denote the domain by* $D^{1,2}$.

Given $u_t(\omega) \in L^2(\Omega, K)$, which is Skorokhod integrable, let

$$u_t(\omega) = \sum_{p=0}^{\infty} I_p(f_p(\cdot, t)).$$

Let us consider $g(s,t) = D_s(u_t(\omega))$. Assume that for each fixed t, $D_s(u_t)$ is well defined and $E\|g(s,t)\|_{K^{\otimes 2}}^2 < \infty$. Suppose

$$f_p(t_1, \ldots, t_p, t) = \sum_{\alpha_1, \ldots, \alpha_{p+1}} a_{\alpha_1, \ldots, \alpha_{p+1}}^p e_{\alpha_1}(t_1) \ldots e_{\alpha_p}(t_p) e_{\alpha_{p+1}}(t),$$

then

$$g(s,t) = D_s(u_t) = \sum_{p=1}^{\infty} p I_{p-1}(f_p(\cdot, s, t)),$$

with

$$I_{p-1}(f_p(\cdot, s, t)) =$$
$$\sum_{\alpha_1, \ldots, \alpha_{p+1}} a_{\alpha_1, \ldots, \alpha_{p+1}}^p I_{p-1}\left(e_{\alpha_1}(t_1) \ldots e_{\alpha_{p-1}}(t_{p-1})\right) e_{\alpha_p}(t_p) e_{\alpha_{p+1}}(t)$$

$$\sum_{\alpha_p, \alpha_{p+1}} \left\{ I_{p-1}\left\{ \sum_{\alpha_1, \ldots, \alpha_{p-1}} a_{\alpha_1, \ldots, \alpha_{p+1}}^p e_{\alpha_1}(t_1) \ldots e_{\alpha_{p-1}}(t_{p-1}) \right\} \right\} e_{\alpha_p}(t_p) e_{\alpha_{p+1}}(t).$$

Since

$$E \sum_{\alpha_p, \alpha_{p+1}} \left\{ I_{p-1}\left\{ \sum_{\alpha_1, \ldots, \alpha_{p-1}} a_{\alpha_1, \ldots, \alpha_{p+1}}^p e_{\alpha_1}(t_1) \ldots e_{\alpha_{p-1}}(t_{p-1}) e_{\alpha_p}(t_p) e_{\alpha_{p+1}}(t) \right\} \right\}^2$$

$$= \sum_{\alpha_p, \alpha_{p+1}} (p-1)! \sum_{\alpha_1, \ldots, \alpha_{p-1}} \left(a_{\alpha_1, \ldots, \alpha_{p+1}}^p \right)^2$$

$$= (p-1)! \sum_{\alpha_1, \ldots, \alpha_{p+1}} \left(a_{\alpha_1, \ldots, \alpha_{p+1}}^p \right)^2$$

$$= (p-1)! \left\| f_p \right\|_{K^{\otimes p+1}}^2 < \infty,$$

we obtain that P–a.s.

$$\sum_{\alpha_p, \alpha_{p+1}} \left\{ I_{p-1} \left\{ \sum_{\alpha_1, \dots, \alpha_{p-1}} a_{\alpha_1, \dots, \alpha_{p+1}}^p e_{\alpha_1}(t_1) \dots e_{\alpha_{p-1}}(t_{p-1}) e_{\alpha_p}(t_p) e_{\alpha_{p+1}}(t) \right\} \right\}^2 < \infty.$$

Hence, for fixed ω, $I_{p-1}\left(f_p(\cdot, s, t) \right)(\omega) \in K^{\otimes 2}$.

Exercise 3.3.1. *Show that*

$$E \left\langle I_{p-1}\left(f_p(\cdot, s, t) \right), I_{q-1}\left(f_q(\cdot, s, t) \right) \right\rangle_{K^{\otimes 2}} = \begin{cases} (p-1)! \langle f_p, f_q \rangle_{K^{\otimes p+1}} & \text{if } p = q \\ 0 & \text{if } p \neq q \end{cases}$$

From the result of Exercise 3.3.1, we conclude that

$$\begin{aligned}
E \left\| g(s, t) \right\|_{K^{\otimes 2}}^2 &= E \left(\sum_{p=1}^{\infty} p I_{p-1}\left(f_p(\cdot, s, t) \right) \right)^2 \\
&= \sum_{p=1}^{\infty} p^2 E \left\| I_{p-1}\left(f_p(\cdot, s, t) \right) \right\|_{K^{\otimes 2}}^2 \\
&= \sum_{p=1}^{\infty} p^2 (p-1)! \left\| f_p \right\|_{K^{\otimes p+1}}^2 \\
&= \sum_{p=1}^{\infty} p \, p! \left\| f_p \right\|_{K^{\otimes p+1}}^2.
\end{aligned}$$

Now let us compute

$$\begin{aligned}
& E \left\langle g(s, t), g(t, s) \right\rangle_{K^{\otimes 2}} \\
&= E \left\langle \sum_{p=1}^{\infty} p I_{p-1}\left(f_p(\cdot, s, t) \right), \sum_{p=1}^{\infty} p I_{p-1}\left(f_p(\cdot, t, s) \right) \right\rangle_{K^{\otimes 2}} \\
&= \sum_{p=1}^{\infty} p^2 E \left\langle I_{p-1}\left(f_p(\cdot, s, t) \right), I_{p-1}\left(f_p(\cdot, t, s) \right) \right\rangle_{K^{\otimes 2}}.
\end{aligned}$$

But

$$\begin{aligned}
& E \left\langle I_{p-1}\left(f_p(\cdot, s, t) \right), I_{p-1}\left(f_p(\cdot, t, s) \right) \right\rangle_{K^{\otimes 2}} \\
&= \sum_{\alpha_p = \alpha_{p+1}} (p-1)! \sum_{\alpha_1, \dots, \alpha_{p-1}} \left(a_{\alpha_1, \dots, \alpha_{p+1}}^p \right)^2 \\
&= (p-1)! \sum_{\alpha_p = \alpha_{p+1}} \sum_{\alpha_1, \dots, \alpha_{p-1}} \left(a_{\alpha_1, \dots, \alpha_{p+1}}^p \right)^2,
\end{aligned}$$

giving

$$E \langle g(s,t), g(t,s) \rangle_{K^{\otimes 2}} = \sum_{p=1}^{\infty} p^2(p-1)! \sum_{\alpha_p = \alpha_{p+1}} \sum_{\alpha_1,\dots,\alpha_{p-1}} \left(a^p_{\alpha_1,\dots,\alpha_{p+1}} \right)^2.$$

Note that

$$\left\| \tilde{f}_p \right\|^2_{K^{\otimes p+1}} = \left\| \frac{1}{p+1} \sum_{\pi} f_p(\pi \underline{t}) \right\|^2_{K^{\otimes p+1}},$$

where $\underline{t} = (t_1, \dots, t_{p+1})$, and the sum is taken over all permutations π which only permute t_{p+1} with one of the t_i's. Hence

$$
\begin{aligned}
\left\| \tilde{f}_p \right\|^2_{K^{\otimes p+1}} &= \frac{1}{(p+1)^2} \sum_{\pi,\pi'} \langle f_p(\pi \underline{t}), f_p(\pi' \underline{t}) \rangle^2_{K^{\otimes p+1}} \\
&= \frac{1}{(p+1)^2} \sum_{\pi} \langle f_p(\pi \underline{t}), f_p(\pi \underline{t}) \rangle^2_{K^{\otimes p+1}} \\
&\quad + \frac{1}{(p+1)^2} \sum_{\pi \ne \pi'} \langle f_p(\pi \underline{t}), f_p(\pi' \underline{t}) \rangle^2_{K^{\otimes p+1}}.
\end{aligned}
$$

The first sum has $p+1$ terms, which are equal to $\left\| f_p \right\|^2_{K^{\otimes p+1}}$. The second sum has $(p+1)p$ terms, which are of the form

$$\sum_{\alpha_i = \alpha_{p+1}} \sum_{\alpha_1,\dots,\alpha_{i-1},\alpha_{i+1},\dots,\alpha_p} \left(a^p_{\alpha_1,\dots,\alpha_{p+1}} \right)^2, \quad (i < p+1).$$

But all these terms are the same because f_p is symmetric in the first p variables, hence the second sum equals

$$(p+1)p \sum_{\alpha_p = \alpha_{p+1}} \sum_{\alpha_1,\dots,\alpha_{p-1}} \left(a^p_{\alpha_1,\dots,\alpha_{p+1}} \right)^2.$$

Therefore,

$$\left\| \tilde{f}_p \right\|^2_{K^{\otimes p+1}} = \frac{1}{p+1} \left\| f_p \right\|^2_{K^{\otimes p+1}} + \frac{p}{p+1} \sum_{\alpha_p = \alpha_{p+1}} \sum_{\alpha_1,\dots,\alpha_{p-1}} \left(a^p_{\alpha_1,\dots,\alpha_{p+1}} \right)^2.$$

Also,

$$
\begin{aligned}
E \left(I^s(u) \right)^2 &= \sum_{p=0}^{\infty} (p+1)! \left\| \tilde{f}_p \right\|^2_{K^{\otimes p+1}} \\
&= \sum_{p=0}^{\infty} (p+1)! \frac{1}{p+1} \left\| f_p \right\|^2_{K^{\otimes p+1}}
\end{aligned}
$$

$$+ \sum_{p=1}^{\infty} (p+1)! \frac{p}{p+1} \sum_{\alpha_p = \alpha_{p+1}} \sum_{\alpha_1,...,\alpha_{p-1}} \left(a^p_{\alpha_1,...,\alpha_{p+1}} \right)^2$$

$$= \sum_{p=0}^{\infty} p! \left\| f_p \right\|^2_{K^{\otimes p+1}} + \sum_{p=1}^{\infty} pp! \sum_{\alpha_p = \alpha_{p+1}} \sum_{\alpha_1,...,\alpha_{p-1}} \left(a^p_{\alpha_1,...,\alpha_{p+1}} \right)^2$$

$$= E \left\| u \right\|^2_K + E \langle g(s,t), g(t,s) \rangle_{K^{\otimes 2}}$$

$$= E \left\| u \right\|^2_K + E \langle D_s(u_t), D_t(u_s) \rangle_{K^{\otimes 2}}.$$

We summarize the above calculations in the following theorem.

Theorem 3.3.1. *Let $u \in L^2(\Omega, K)$. Assume that for each fixed t, $D_s(u_t)$ is well defined and $E \left\| D_s(u_t) \right\|^2_{K^{\otimes 2}} < \infty$. Then u is Skorokhod integrable and*

$$E \left(I^s(u) \right)^2 = E \left\| u \right\|^2_K + E \langle D_s(u_t), D_t(u_s) \rangle_{K^{\otimes 2}}. \tag{3.17}$$

We now provide examples of Skorokhod integrals.

Example 3.3.1. *We discuss examples of Skorokhod integrals in specific cases.*

(a) *Brownian Motion.*

Let us consider $C(t,s) = t \wedge s$, $s, t \in [0,1]$, the covariance function for Brownian motion W. We know from Example 1.2.1 that the RKHS is

$$K(C) = \left\{ g \mid g(t) = \int_0^t f_g(s) \, ds, \ f_g \in L^2([0,1]) \right\},$$

and the scalar product in K is given by

$$\langle g_1, g_2 \rangle_K = \langle f_{g_1}, f_{g_2} \rangle_{L^2([0,1])}.$$

Define $V : L^2([0,1]) \to K = K(C)$ by

$$V(f) = \int_0^\cdot f(s) \, ds.$$

Using Theorem 2.2.1, we define

$$\pi(V(f)) = \int_0^1 f(s) \, dW_s.$$

The multiple Wiener integral $I_p(g_p)$ can now be defined for $g_p \in K^{\odot p}$ as in Definition 3.1.1, so that

$$I_p \left(e_1^{p_1} ... e_n^{p_n} \right) = \prod_{i=1}^{n} H_{p_i} \left(\pi(e_i) \right),$$

where

$$e_1^{p_1} = e_1(t_1)...e_1\left(t_{p_1}\right)$$

$$e_2^{p_2} \;=\; e_2\left(t_{p_1+1}\right)...e_2\left(t_{p_1+p_2}\right)$$

$$...$$

$$e_n^{p_n} \;=\; e_n\left(t_{p_1+...+p_{n-1}+1}\right)...e_n\left(t_{p_1+...+p_n}\right).$$

and $e_1,...,e_n$ are orthonormal in K. If $e_i = V(f_i)$, then $f_1,...,f_n$ are orthonormal in $L^2([0,1])$. Hence, if we define

$$I_p^i\left(f_1^{p_1}...f_2^{p_2}\right) = \prod_{i=1}^{n} H_{p_i}\left(\pi\left(V(f_i)\right)\right),$$

then I_p^i can be uniquely extended to $L^2([0,1])^{\odot p}$ because it is isometric to $K^{\odot p}$. But $L^2([0,1])^{\odot p} = L^2([0,1]^p)^{\wedge}$, the space of symmetric functions in $L^2([0,1]^p)$. We leave it to the reader to check that for elementary functions this definition coincides with the definition in [48]. Since $I_p\left(K^{\odot p}\right) = I_p^i\left(L^2([0,1]^p)^{\wedge}\right)$, we have

$$L^2\left(\Omega, \sigma\{W_t, t \in [0,1]\}, P\right) = \bigoplus_{p=0}^{\infty} I_p^i\left(L^2([0,1]^p)^{\wedge}\right).$$

Thus the definition of the Skorokhod integral I^s given in Definition 3.2.1 coincides with the one given in [94] and [95]. Hence, we have the following theorem (Proposition 1.3.11 in [95]).

Theorem 3.3.2. *Adapted processes in $L^2([0,1] \times \Omega)$ are Skorokhod integrable and for such integrands, the Skorokhod and Itô integrals with respect to Brownian motion coincide.*

We will see in Section 4.1.2 that the Skorokhod and Malliavin derivatives also coincide.

(b) *Multidimensional Brownian Motion.*
Consider a Gaussian process $\{W_i(t), (i,t) \in \{1,...,d\} \times [0,1]\}$ with covariance $C((i,t),(j,s)) = \delta_{i,j}(t \wedge s)$. Then clearly the RKHS is as follows:

$$K(C) = \left\{ g \,\big|\, g_i(t) = \int_0^t f_i(s)\,ds,\; f_i \in L^2([0,1]),\; i \in \{1,...,d\} \right\}.$$

We can identify $K(C)$ with $\bigoplus_{i=1}^{d} K(C')$, where $C'(t,s) = t \wedge s$. Hence, there exists an isometry

$$V : \bigoplus_{i=1}^{d} L^2([0,1]) \to K(C).$$

Now, as in [80], the stochastic integral π takes the form

$$\pi(V(f)) = \sum_{i=1}^{d} \int_0^1 f_i(s)\,dW_i(s).$$

By adopting the arguments in part (a) of this example, we can see that the corresponding multiple Wiener integrals coincide with those defined in [96], and hence the Skorokhod integrals are the same. The details are left for the reader as an exercise.

One can also consider cylindrical Brownian motion (or cylindrical Wiener process, see Section 2.1.2 in [36]) $W(t,h)$ with covariance defined on $H \times [0,1]$ by $C((h,t),(g,s)) = \langle h,g \rangle_H t \wedge s$. In this case, $K(C) = H \otimes K(C')$ and one can define for $h \otimes f \in K(C)$

$$\pi(h \otimes f) = I_1(h \otimes f) = \sum_{i=1}^{\infty} \langle h,h_i \rangle_H \int_0^1 f(s)\, dW(e_i,s),$$

where $\{h_i\}_{i=1}^{\infty}$ is an orthonormal basis in H. The development of multiple Wiener integrals and of the Skorokhod integral are left to the reader as an exercise.

(c) *Random Linear Functional on a Hilbert Space H [118].*

Consider a random linear functional ξ on a separable Hilbert space H into a Gaussian space G, continuous in the norm of H, such that

$$\langle \xi(h), \xi(h') \rangle_G = \langle h,h' \rangle_H.$$

The Gaussian process $\{\xi(h),\ h \in H\}$ has covariance $C(h,h') = \langle h,h' \rangle_H$. In view of the Riesz Representation Theorem, we can identify elements of $K(C) = K$ with elements of H. Hence, the Skorokhod integral of a K-valued random variable η can be identified with Skorokhod integral of an H-valued random variable, denoted by the same symbol η, with $E\|\eta\|_H^2 < \infty$. This is precisely the class of integrable random variables considered by Skorokhod in [118]. To compare Definition 3.2.1 to that of Skorokhod, note that $K^{\otimes p} = K(C)^{\otimes p}$ has reproducing kernel

$$C(h_1,g_1)...C(h_p,g_p) = \langle h_1,g_1 \rangle_H ... \langle h_p,g_p \rangle_H,$$

and hence any $f \in K(C)^{\otimes p}$ satisfies

$$f(h_1,...,h_p) = \langle f, C(h_1,\cdot)...C(h_p,\cdot) \rangle_{K^{\otimes p}}.$$

This relation shows that elements f of $K^{\otimes p}$ are symmetric multilinear forms and

$$f(h_1,...,h_p) = \sum_{\alpha_1,...,\alpha_p} f(e_{\alpha_1}(h_1)...e_{\alpha_p}(h_p))$$

with $\{e_\alpha,\ \alpha \in J\}$ an orthonormal basis in $K(C) \cong H$. Now $f \in K^{\odot p}$ implies that $f(e_{\alpha_1},...,e_{\alpha_p})$ is symmetric in $(\alpha_1,...,\alpha_p)$ and

$$\sum_{\alpha_1,...,\alpha_p} \left(f(e_{\alpha_1},...,e_{\alpha_p}) \right)^2 < \infty.$$

Thus, I_p is defined on the class of p-linear operators A on $H^{\otimes p}$ with $\text{tr}(A^*A)^p < \infty$ *in the sense of [118]. As in part (a) of this example,*

$$I_p\left(e_1^{p_1}...e_n^{p_n}\right) = \prod_{i=1}^{n} H_{p_i}\left(\langle\xi,e_i\rangle\right)$$

and since $\pi(x) = I_1(x) = \langle\cdot,x\rangle$, the definition of I_p in [118] coincides with the I_p defined above.

Exercise 3.3.2. *Verify that for elementary functions the definition of multiple Wiener integral in part (a) of Example 3.3.1 coincides with the definition in [94]. Verify the multiple Wiener integrals defined in part (b) of Example 3.3.1 coincide with those defined in [96]. Complete the development of multiple Wiener integrals and the Skorokhod integral for the case of cylindrical Wiener process.*

3.4 Ogawa Integral

We now define the Ogawa integral, [98],[99],[100], with respect to any centered Gaussian random field. Let $g: \Omega \to K$ be a K-valued Bochner measurable function. Hence, $g.(\omega)$ takes values in a separable subspace of K. Let $\{e_i\}_{i=1}^{\infty}$ be an orthonormal basis of this subspace.

Definition 3.4.1. *Let T be any set and $\{X_t, t \in T\}$ be a centered Gaussian random field with covariance C and $K = K(C)$ be the associated RKHS. Let $g: \Omega \to K$ be a K-valued Bochner measurable function.*

(a) *A K-valued Bochner measurable function g is said to be Ogawa integrable with respect to an orthonormal basis $\{e_i\}_{i=1}^{\infty}$ if the following series converges in probability,*

$$\sum_{n=1}^{\infty} \langle g,e_n\rangle_K I_1\left(e_n\right) \tag{3.18}$$

In this case, the limit is called Ogawa integral of g and we denote it by $\delta^e(g)$.

(b) *If the limit in part (a) exists with respect to all orthonormal bases and is independent of the basis, then the function g is called universaly Ogawa integrable and the integral is called universal Ogawa integral and we denote it by $\delta^o(g)$.*

Now we relate the Ogawa ond Skorokhod integrals when they exist. We know from Theorem 3.3.1 that $g \in L_2(\Omega,K)$ is Skorohod integrable if $E\|D_*g.\|_{K^{\otimes 2}}^2 < \infty$ where D_* is the Skorohod derivative. We begin with a supporting lemma.

Lemma 3.4.1. *Let $F \in L_2(\Omega,\mathcal{F},P)$, $D_*F \in L_2(\Omega,K)$ and $f \in K$. Then Ff is Skorohod integrable and $I^s(Ff) = I_1(f)F - \langle D_*F, f(*)\rangle_K$.*

Proof. Let $F = I_p(f_p)$, with $f_p \in K^{\odot p}$. Then $Ff(t) = I_p(f_p(\cdot)f(t))$. Clearly, by (3.14), Ff is Skorohod integrable and using (3.13) and (3.7) we have

$$
\begin{aligned}
I^s(Ff) = I_{p+1}\left(\widetilde{f_p(\cdot)f(*)}\right) &= I_{p+1}\left(f_p(\cdot)f(*)\right) \\
&= I_p\left(f_p(\cdot)\right)I_1(f(*)) - \\
&\quad pI_{p-1}\left(\langle f_p(t_1,\ldots,t_{p-1},\cdot), f(\cdot)\rangle_K\right).
\end{aligned}
$$

Therefore,

$$
I^s(Ff) = FI^s(f) - \langle D_*F, f(*)\rangle_K.
$$

Suppose $F = \sum_{p=0}^{\infty} I_p(f_p)$, and consider

$$
\begin{aligned}
\sum_{p=0}^{\infty}(p+1)! &\left\| \frac{1}{p+1}\left(f_p(t_1,\ldots,t_p)f(t) \right.\right. \\
&\left.\left. + \sum_{i=1}^{p} f_p(t_1,\ldots t_{i-1},t,t_{i+1},\ldots,t_p)f(t_i) \right) \right\|^2_{K^{\otimes p+1}} \\
&\leq \sum_{p=0}^{\infty} \frac{(p+1)!}{(p+1)^2}(p+1)^2 \|f_pf\|^2_{K^{\otimes p+1}} \\
&= \|f\|^2_K \left(\sum_{p=0}^{\infty} p! \|f_p\|^2_{K^{\otimes p}} + \sum p!p \|f_p\|^2_{K^{\otimes p}} \right) \\
&\leq \|f\|^2_K \left(E(F)^2 + E\|D_*F\|^2_K \right),
\end{aligned}
$$

that is, Ff is Skorohod integrable and

$$
\begin{aligned}
I^s(Ff) &= \lim_{N\to\infty} \sum_{p=0}^{N} I^s\left(I_p(f_p)f\right) \\
&= \lim_{N\to\infty} \left(\sum_{p=0}^{N} I_p(f_p)I^s(f) \right. \\
&\quad \left. - \sum_{p=0}^{N} pI_{p-1}\left(\langle f_p(t_1,\ldots,t_{p-1},\cdot), f(\cdot)\rangle_K\right) \right)
\end{aligned}
$$

Since $E\left(\langle D_*F, f(*)\rangle^2_K \right) < \infty$ and Ff is Skorohod integrable, both terms on the RHS converge in $L_2(\Omega)$. $\square$

Let $\{e_n\}_{n=1}^{\infty}$ be an ONB of K and $u \in L_2(\Omega, K)$. Define

$$
\tau^e(Du) = \sum_{n=1}^{\infty} \langle D_* \langle u., e_n(\cdot)\rangle_K, e_n(*)\rangle_K,
$$

where the limit is in probability.

The Ogawa and Skorokhod integrals are related in the following theorem.

Theorem 3.4.1. *Let $u \in L^2(\Omega, K)$ be Skorohod integrable and for every $h \in K$, $D_*\langle u, h \rangle_K \in L^2(\Omega, K)$.*

(a) *Then u is Ogawa integrable with respect to an orthonolmal basis $\{e_n\}_{n=1}^{\infty} \subseteq K$ if and only if $\tau^e(Du)$ exists.*

(b) *In addition,*

$$\sum_{n=1}^{\infty} I_1(e_n) \langle u, e_n \rangle_K \to \delta^e(u)$$

in $L^2(\Omega)$ if and only if $\tau^e(Du)$ exists and the series defining it converges in $L^2(\Omega)$.

In both cases,

$$\delta^e(u) = I^s(u) + \tau^e(Du)$$

Proof. By Lemma 3.4.1,

$$I^s\left(\sum_{n=1}^{N} \langle u., e_n(\cdot)\rangle_K e_n(\cdot)\right)$$

$$= \sum_{n=1}^{N} I^s\left(\langle u., e_n(\cdot)\rangle_K e_n(\cdot)\right)$$

$$= \sum_{n=1}^{N} I_1(e_n) \langle u., e_n(\cdot)\rangle_K - \sum_{n=1}^{N} \langle D_* \langle u., e_n(\cdot)\rangle_K, e_n(*)\rangle_K.$$

$\square$

Corollary 3.4.1. *Let $u \in L_2(\Omega, K)$ be of the form $u_t = \sum_{p=0}^{\infty} I_p(f_p(\cdot, t))$. Then the condition*

$$\sum_{p=0}^{\infty} (p+1)! \left\| \tilde{f}_p(\cdot, *) \right\|_{K^{\otimes p+1}}^2 + \sum_{p=1}^{\infty} pp! \left\| \tau^e(f_p) \right\|_{K^{\otimes p-1}}^2 < \infty \qquad (3.19)$$

implies that the Ogawa integral $\delta^e(u)$ exists as a limit in $L_2(\Omega)$ and $\delta^e(u) = I^s(u) + \tau^e(Du)$, where

$$\tau^e(f_p)(\cdot) = \sum_{n=1}^{\infty} \langle \langle f_p(\cdot, s, t), e_n(t)\rangle_K, e_n(s)\rangle_K.$$

Proof. If u has a decomposition

$$u_t = \sum_{p=0}^{\infty} I_p(f_p(\cdot, t)) \qquad \text{for } t \in T,$$

then

$$\langle u., e_n(\cdot)\rangle_K = \left\langle \sum_{p=0}^{\infty} I_p(f_p(\cdot, t)), e_n(t) \right\rangle_K$$

$$= \sum_{p=0}^{\infty} \langle I_p\big(f_p(\cdot,t)\big), e_n(t)\rangle_K.$$

But

$$\langle I_p\big(f_p(\cdot,t)\big), e_n(t)\rangle_K = I_p\big(\langle f_p(\cdot,t), e_n(t)\rangle_K\big).$$

Therefore, we have

$$D_s\langle u., e_n(\cdot)\rangle_K = \sum_{p=1}^{\infty} p I_{p-1}\big(\langle f_p(\cdot,s,t), e_n(t)\rangle_K\big),$$

giving

$$
\begin{aligned}
\tau^e(Du) &= \sum_{n=1}^{\infty} \Big\langle \sum_{p=1}^{\infty} p I_{p-1}\big(\langle f_p(\cdot,s,t), e_n(t)\rangle_K\big), e_n(s)\Big\rangle_K \\
&= \sum_{n=1}^{\infty}\sum_{p=1}^{\infty} p I_{p-1}\big(\langle \langle f_p(\cdot,s,t), e_n(t)\rangle_K, e_n(s)\rangle_K\big).
\end{aligned}
$$

The series in (3.19) are the $L^2(\Omega)$ norms of $I^s(u)$ and $\tau^e(Du)$, respectively.
□

To relate the universal Ogawa and Skorokhod integrals, we follow the work in [109] and [31]. Let $f_p(\cdot,s,t) \in K^{\otimes p+1}$ be such that $f_p(\cdot,t) \in K^{\odot p}$ for all $t \in T$. For $p \geq 1$, define $F_p: K \to K^{\otimes p}$ by

$$(F_p\varphi)(t) = \langle \varphi(*), f_p(\cdot,*,t)\rangle_K.$$

Assume that F_p has a *summable trace*, which means that for all orthonormal bases $\{e_n\}_{n=1}^{\infty} \subseteq K$ the series

$$\sum_{n=1}^{\infty} \langle \langle e_n(s), f_p(\cdot,s,t)\rangle_K, e_n(t)\rangle_K \qquad (3.20)$$

exists in $K^{\otimes p-1}$. We have the following fact about F_p (Corollary 2.2 in [109]).

Proposition 3.4.1. *If F_p has a summable trace, then the series in (3.20) does not depend on the choice of the orthonormal basis $\{e_n\}_{n=1}^{\infty}$.*

Proof. For any $h \in K^{\otimes p-1}$ consider the operator $\tilde{F}_{p,h}: K \to K$ defined by

$$(\tilde{F}_{p,h}k)(t) = \Big\langle \langle \tfrac{1}{2}\big(f_p(\cdot,s,t)+f_p(\cdot,t,s)\big), h(\cdot)\rangle_{K^{\otimes p-1}}, k(s)\Big\rangle_K. \qquad (3.21)$$

Denote

$$k_n = \langle \langle e_n(s), f_p(\cdot,s,t)\rangle_K, e_n(t)\rangle_K.$$

Since F_p has summable trace, the sum $\sum_{n=1}^{\infty} k_n$ converges in $K^{\otimes p-1}$. In fact, since any permutation of the basis $\{e_n\}_{n=1}^{\infty}$ is again an orthonormal basis, we conclude that this convergence is unconditional. Hence, for any $h \in K$,

$$\sum_{n=1}^{\infty} \left| \left(\left(\tilde{F}_{p,h} e_n \right)(\cdot), e_n(\cdot) \right)_K \right| = \sum_{n=1}^{\infty} |\langle k_n, h \rangle_K| < \infty. \tag{3.22}$$

The operator $\tilde{F}_{p,h}$ is also self-adjoint. From the proof of Theorem 3.4.3 in [4] (see also Exercise 3.4.1) we conclude that $\tilde{F}_{p,h}$ is a trace-class operator on K. Therefore, one can define a linear functional on $K^{\otimes p-1}$,

$$
\begin{aligned}
h \quad &\rightarrow \quad \mathrm{tr}\tilde{F}_{p,h} \\
&= \quad \sum_{n=1}^{\infty} \left\langle \tilde{F}_{p,h} e_n, e_n \right\rangle_K \\
&= \quad \sum_{n=1}^{\infty} \left\langle \left\langle \left\langle \frac{f_p(\cdot,s,t) + f_p(\cdot,t,s)}{2}, h(\cdot) \right\rangle_{K^{\otimes p-1}}, e_n(s) \right\rangle_K, e_n(t) \right\rangle_K \\
&= \quad \sum_{n=1}^{\infty} \left\langle \left\langle \left\langle f_p(\cdot,s,t), e_n(s) \right\rangle_K, e_n(t) \right\rangle_K, h(\cdot) \right\rangle_{K^{\otimes p-1}} \\
&= \quad \langle k_0, h \rangle_K,
\end{aligned}
$$

for some $k_0 \in K^{\otimes p-1}$ by the Riesz theorem. Hence, the sum in (3.20) equals h_0 regardless of the choice of the orthonormal basis in K.   □

Exercise 3.4.1. *Let K be a separable Hilbert space. Show that if $T \in L(K)$ is self-adjoint and satisfies*

$$\sum_{n=1}^{\infty} |\langle T e_n, e_n \rangle_K| < \infty$$

for all orthonormal bases $\{e_n\}_{n=1}^{\infty} \subseteq K$, then T is a trace-class operator on K.

Exercise 3.4.2. *Show that the operator F_p has summable trace if and only if the operator $\tilde{F}_{p,h}$ is trace class for every $h \in K$.*

Using the result of Proposition 3.4.1 we define

$$\tau(f_p) = \sum_{n=1}^{\infty} \left\langle \left\langle f_p(\cdot,s,t), e_n(s) \right\rangle_K, e_n(t) \right\rangle_K \tag{3.23}$$

whenever F_p has a summable trace, so that $\tau(f_p)$ is well defined.

Corollary 3.4.2. *Let u be as in Corollary 3.4.1. If for every $p \geq 1$, F_p has a summable trace and*

$$\sum_{p=0}^{\infty} (p+1)! \left\| \tilde{f}_p(\cdot,*) \right\|_{K^{\otimes p+1}}^2 + \sum_{p=1}^{\infty} p\, p! \left\| \tau(f_p) \right\|_{K^{\otimes p-1}}^2 < \infty, \tag{3.24}$$

then the universal Ogawa integral of u exists and

$$\delta^o(u) = I^s(u) + \sum_{p=1}^{\infty} p I_{p-1}\left(\tau(f_p) \right).$$

Let $T \in \mathcal{L}(K)$, recall the *trace norm* of T,

$$\|T\|_\tau = \mathrm{tr}\left((T^*T)^{1/2}\right).$$

Exercise 3.4.3. *Show that for F_p of a summable trace*

$$\left\|\tau\left(f_p\right)\right\|_{K^{\otimes p-1}} = \sup\left\{\left\|\tilde{F}_{p,h}\right\|_\tau \mid h \in K^{\otimes p-1}, \|h\|_{K^{\otimes p-1}} \le 1\right\}.$$

Prove that F_p is of a summable trace if and only if $\left\|\tau\left(f_p\right)\right\|_{K^{\otimes p-1}} < \infty.$

Remark 3.4.1. *An upper bound on $\left\|\tau\left(f_p\right)\right\|_{K^{\otimes p-1}}$ is provided in [109].*

The following result, Theorem 3.3 in [109], helps determine if the universal Ogawa integral does not exist.

Theorem 3.4.2. *Let $u_t = I_p(f(\cdot,t))$, $t \in T$, where $f(\cdot,t) \in K^{\otimes p+1}$ and for each t, $f(\cdot,t) \in K^{\odot p}$. Then $\delta^o(u)$ exists as a limit in $L^2(\Omega)$ if and only if the map $F: K \to K^{\otimes p-1}$ defined by*

$$(Fk)(t) = \langle k(s), f(\cdot,s,t)\rangle_K \tag{3.25}$$

has a summable trace. In this case

$$\delta^o(u) = I_{p+1}(f) + pI_{p-1}\tau(f),$$

where $\tau(f)$ is defined in (3.23).

Example 3.4.1. **(a)** *Brownian motion.*
Let $\{B_t, t \in [0,1]\}$ be the standard Brownian motion of Example 1.2.1. Then the process $u_t = \int_0^t B_s ds$ is an H-valued stochastic process, where H is the RKHS of Brownian motion. The Ogawa integral of u is given by

$$\delta^o(u) = \sum_{n=1}^\infty (u,e_n)\pi(e_n) = \sum_{n=1}^\infty (B,e_n')_{L_2([0,1])} \int_0^1 e_n' dB = \frac{1}{2}B_1^2 \quad \left(e_n' = \frac{de_n}{dt}\right)$$

as proved in Example 3.2 [109] (the above formula is correct for any ONB $\{e_n\}_{n=1}^\infty$ in H).

(b) *Reversed Brownian motion.*
Let us now consider the reversed Brownian motion process B_{1-t}. The Skorohod integral $I^s(u)$ exists. The Skorokhod derivative of u is given by $Du(h) = \int_0^\cdot h(1-s)ds$.
Example in [109] shows that u is not universally Ogawa integrable in the sense of [109]. Note that convergence in [109] is in $L_2(\Omega)$

(c) *Ogawa non-integrable process.*
The following example in [99] shows that given any ONB in H one can construct a process that is not Ogawa integrable with respect to this basis. Let

$$u_t = \sum_{n=1}^\infty \frac{1}{n^p} e_n(t) \mathrm{sign}(\pi(e_n)) \quad (1/2 < p \le 1),$$

then $u \in L_2(\Omega,H)$, but the series defining $\delta^o(u)$,

$$\sum_{n=1}^{\infty} \frac{1}{n^p} \operatorname{sign}(\pi(e_n)) \pi(e_n) = \sum_{n=1}^{\infty} \frac{1}{n^p} |\pi(e_n)|$$

diverges a.e.

(d) *An example of a process with an infinite Wiener Chaos expansion.*
Consider the general case of a Gaussian process $X = \{X_t, \, t \in T\}$ defined on a probability space $(\Omega,\mathcal{F},P)$ and the associated triple $(\iota,H,\mathcal{X})$. Let

$$
\begin{aligned}
u_t &= \sum_{n=1}^{\infty} \frac{1}{n^p} \frac{1}{\sqrt{n!}} I_n(e_1(t_1)...e_1(t_n)) e_n(t) \\
&= \sum_{n=1}^{\infty} \frac{1}{n^p} \frac{1}{\sqrt{n!}} \mathcal{H}_n(\pi(e_1)) e_n(t)
\end{aligned}
$$

where $\{e_n\}_{n=1}^{\infty} \subset H$ is an ONB of H, $\mathcal{H}_n$'s are Hermite polynomials normalized as in Section 2.1.

(1) $u \in L_2(\Omega,H)$ *if and only if $p > 1/2$, since*

$$E\|u\|_H^2 = E \sum_{n=1}^{\infty} \frac{1}{n^{2p}} \frac{1}{n!} \mathcal{H}_n^2(\pi(e_1)) = \sum_{n=1}^{\infty} \frac{1}{n^{2p}}.$$

(2) *u is Ogawa integrable with respect to ONB $\{e_n\}_{n=1}^{\infty}$ if $p > 1/2$. Actually u is universally Ogawa integrable as we have*

$$\sum_{n=1}^{\infty} \langle u, e_n \rangle_H \pi(e_n) = \sum_{n=1}^{\infty} \frac{1}{n^p} \frac{1}{\sqrt{n!}} \mathcal{H}_n(\pi(e_1)) \pi(e_n).$$

We need to check when this series converges in probability. Since (excluding the first term) the series consists of centered, integrable random variables adapted to the filtration $\mathcal{F}_n = \sigma\{I_1(e_1), I_1(e_2), ..., I_1(e_n)\}$, $n \geq 1$, $\mathcal{F}_0 = \{\varnothing, \Omega\}$, it converges P–a.e. on the following set (see [91]):

$$
\begin{aligned}
\Omega_0 &= \left\{ \sum_{n=1}^{\infty} E\left(\left(\frac{1}{n^p} \frac{1}{\sqrt{n!}} \mathcal{H}_n(I_1(e_1)) I_1(e_n) \right)^2 \Big| \mathcal{F}_{n-1} \right) < \infty \right\} \\
&= \left\{ \sum_{n=1}^{\infty} \frac{1}{n^{2p}} \frac{1}{n!} \mathcal{H}_n^2(I_1(e_1)) < \infty \right\}.
\end{aligned}
$$

But

$$E\left(\sum_{n=1}^{\infty} \frac{1}{n^{2p}} \frac{1}{n!} \mathcal{H}_n^2(I_1(e_1)) \right) = \sum_{n=1}^{\infty} \frac{1}{n^{2p}} < \infty,$$

therefore $P(\Omega_0) = 1$ and

$$\delta^e(u) = \sum_{n=1}^{\infty} \frac{1}{n^p} \frac{1}{\sqrt{n!}} \mathcal{H}_n(I_1(e_1)) I_1(e_n).$$

(3) $u \in \mathcal{D}(I^s)$ if and only if $p > 1/2$. We need to show $L_2(\Omega)$ convergence of the series defining the Skorohod integral of u. This can be proved as follows:

$$\sum_{n=1}^{\infty} (n+1)! \left\| \frac{1}{n^p} \frac{1}{\sqrt{n!}} \left(e_1(t_1) \widetilde{\ldots e_1(t_n)} e_n(t) \right) \right\|^2_{H^{\otimes(n+1)}}$$

$$= \sum_{n=1}^{\infty} \frac{n+1}{n^{2p}} \left\| \frac{1}{n+1} \left(e_1(t_1) \ldots e_1(t_n) e_n(t) \right. \right.$$

$$\left. \left. + \sum_{i=1}^{n} e_1(t_1) \ldots e_1(t_{i-1}) e_1(t) e_1(t_{i+1}) \ldots e_1(t_n) e_n(t_i) \right) \right\|^2_{H^{\otimes(n+1)}}$$

$$= 2 + \sum_{n=2}^{\infty} \frac{1}{n^{2p}} \|e_1 \ldots e_1 e_n\|^2_{H^{\otimes(n+1)}} = 2 + \sum_{n=2}^{\infty} \frac{1}{n^{2p}}.$$

Second equality follows from the orthogonality of the components under the norm.
Also, by property (3) of Multiple Wiener Integrals, we get

$$I^s(u) = \sum_{n=1}^{\infty} \frac{1}{n^p} \frac{1}{\sqrt{n!}} \mathcal{H}_n(\pi(e_1)) \pi(e_n) - 1 = \delta^o_e(u) - 1.$$

(4) *Skorokhod derivative.*
  (4.1) $u_t \in \mathcal{D}(D)$ with $Du_t \in L_2(\mathcal{X}, H)$ for t fixed, if $p \geq \frac{1}{2}$.
  For the above to be true, the following series must converge:

$$\sum_{n=1}^{\infty} nn! \left\| \frac{1}{n^p} \frac{1}{\sqrt{n!}} \{e_1 \ldots e_1\} e_n(t) \right\|^2_{H^{\odot n}} = \sum_{n=1}^{\infty} \frac{1}{n^{2p-1}} e_n^2(t) < \infty$$

because $\sum_{n=1}^{\infty} e_n(t)^2 < \infty$ ($e_n(t)$ are the Fourier coefficients of $C(\cdot, t)$ in H).

  (4.2) $u \in \mathcal{D}(D)$ with $Du \in L_2(\Omega, H^{\otimes 2})$ if and only if $p > 1$.
  Indeed

$$\sum_{n=1}^{\infty} nn! \left\| \frac{1}{n^p} \frac{1}{\sqrt{n!}} e_1 \ldots e_1 e_n \right\|^2_{H^{\otimes(n+1)}} = \sum_{n=1}^{\infty} \frac{1}{n^{2p-1}} < \infty.$$

Thus for $\frac{1}{2} < p \leq 1$, $u \in \mathcal{D}(I^s)$ and can be Ogawa integrable with respect to some basis while $D^M u \notin L_2(\mathcal{X}, H^{\otimes 2})$.

3.5   Appendix

Application-Limit Theorems for U-Statistics of Infinite Order
    The concept of special symmetric statistics, called the U-statistics, was introduced by Hoeffding in [44]. Consider a sequence of i.i.d. random variables $X_1, X_2, ..., X_n$ taking values in a measurable space $(\mathcal{X}, \mathcal{B})$, with a common distribution μ. Let h_m be a symmetric function depending on m variables. Then the corresponding U-statistic of order m is defined by

$$U_m^n(h_m) = \frac{1}{\binom{n}{m}} \sigma_m^n(h_m),$$

where

$$\sigma_m^n(h_m) = \begin{cases} \sum_{1 \le s_1 < ... < s_m \le n} h_m\left(X_{s_1}, ..., X_{s_m}\right) & n \ge m \\ 0 & n < m \end{cases}$$

For a symmetric function $h \in L^1\left(\mathcal{X}^m, \mathcal{B}^{\otimes m}, \mu^{\otimes m}\right)$ define

$$Q_i(h) = E^{\mathcal{F}_i}\left(h(X_1, ..., X_n)\right),$$

where $\mathcal{F}_i = \sigma\{X_1, ..., X_{i-1}, X_{i+1}, ..., X_n\}$. Since the transformations $Q_1, ..., Q_n$ commute and

$$h = \left(\left[(I - Q_1) + Q_1\right]...\left[(I - Q_n) + Q_n\right]\right)(h),$$

we have

$$h(X_1, X_2, ..., X_n) = \sum_{k=0}^{\infty} \sigma_k^n(h_k), \qquad (3.26)$$

with

$$h_k(X_1, ..., X_k) = (I - Q_1)...(I - Q_k)Q_{k+1}...Q_n(h).$$

Note that the sum (3.26) is finite and involves $X_{s_1}, ..., X_{s_k}$ with $1 \le s_1, ..., s_k \le n$ for $k \le n$. The functions $h_k(x_1, ..., x_k)$ are symmetric, $k \le n$, and

$$E h_k(x_1, ..., x_{k-1}, X_k) = 0. \qquad (3.27)$$

Following [44], we call (3.26) a *canonical decomposition* and we call the functions h_k present in (3.26) *canonical*. Denote by $\mathcal{S}^n$ the σ-field of symmetric sets in $\mathcal{B}^{\otimes n}$, that is, sets that are invariant under permutations of indexes. Since (3.26) implies that

$$h(X_1, ..., X_n) = \sum_{k=0}^{\infty} \sigma_k^n(\tilde{h}_k),$$

we have

$$
\begin{aligned}
U_k^n(h) &= \binom{n}{m}^{-1} \sigma_m^n(h) = \binom{n}{m}^{-1} \sum_{s_1,\dots,s_m} h\left(X_{s_1},\dots,X_{s_m}\right) \\
&= E\left(h\left(X_{s_1},\dots,X_{s_k}\right)|\mathcal{S}^n\right) = E\left(\sum_{k=0}^{m} \sigma_k^m\left(\tilde{h}_k\right)|\mathcal{S}^n\right) \\
&= \sum_{k=0}^{m} \sigma_k^m\left(E\left(\tilde{h}_k|\mathcal{S}^n\right)\right) = \sum_{k=0}^{m} \sigma_k^m \binom{n}{k}^{-1} \sum_{s_1,\dots,s_k} \tilde{h}_k\left(X_{s_1},\dots,X_{s_k}\right) \\
&= \sum_{k=0}^{m} \binom{n}{k}^{-1}\binom{m}{k} \sum_{s_1,\dots,s_k} \tilde{h}_k\left(X_{s_1},\dots,X_{s_k}\right) = \sum_{k=0}^{m}\binom{n}{k}\sigma_k^n\left(h_k\right) = \sum_{k=0}^{m} U_k^n\left(h_k\right),
\end{aligned}
$$

where $h_k = \binom{n}{k}\tilde{h}_k$. Thus, we obtain Hoeffding's decomposition

$$
U_k^n(h) = \sum_{k=0}^{m} U_k^n\left(h_k\right).
$$

Dynkin and Mandelbaum [22] study limit theorems for

$$
\sum_{k=0}^{\infty} n^{-k/2} \sigma_k^n\left(h_k\right),
$$

with the functions $h_k \in L^2\left(\mathcal{X}^k, \mathcal{B}^{\otimes k}, \mu^{\otimes k}\right)$ being symmetric and satisfying condition (3.27) as a generalization of Hoeffding's work. A simple proof of their result is given by Mandrekar and Rao [82], which we now present here. We define the following space

$$
\mathcal{H} = \left\{ h = (h_0, h_1, \dots, h_k, \dots) : h_k \text{ is canonical and } \sum_{k=0}^{\infty} \frac{\|h_k\|_{L^2(\mathcal{X}^k)}^2}{k!} < \infty \right\},
$$

with a norm given by

$$
\|h\|_{\mathcal{H}}^2 = \sum_{k=0}^{\infty} \frac{\|h_k\|_{L^2(\mathcal{X}^k)}^2}{k!}.
$$

Denote

$$
L_0(\mathcal{X},\mathcal{B},\mu) = \left\{ \psi \in L^2(\mathcal{X},\mathcal{B},\mu) : E\varphi(X_1) = 0 \right\},
$$

and for $\varphi \in L_0(\mathcal{X},\mathcal{B},\mu)$,

$$
h_k^{\varphi}(x_1,\dots,x_k) = \varphi(x_1)\dots\varphi(x_k), \quad k > 0,
$$

and $h_0 \equiv$ constant.

It is easy to check that the linear subspace generated by the set $\left\{ h_k^{\varphi} : \varphi \in L_0(\mathcal{X},\mathcal{B},\mu) \right\}$ is dense in the space of symmetric canonical functions in $L^2(\mathcal{X}^k,\mathcal{B}^{\otimes k},\mu^{\otimes k})$. As a consequence, we obtain that the linear subspace generated by

$$\{h^\varphi = (h_0,h_1,...,h_k,...): \varphi \in L_0(\mathcal{X},\mathcal{B},\mu)\}$$

is dense in $\mathcal{H}$. For $h \in \mathcal{H}$ let us define

$$Y_n(h) = \sum_{k=0}^{\infty} n^{-\frac{k}{2}} \sigma_k^n(h_k).$$

In order to identify the limit we introduce a Gaussian random field W corresponding to the covariance $C_W(A,B) = \mu(A\cup B)$ for $A,B \in \mathcal{B}$. Clearly $W(A)$ and $W(B)$ are independent Gaussian random variables if $A\cup B = \varnothing$. Denote by $K(C_W)$ the RKHS of C_W. Then we know from part (f) of Example 1.2.1 that

$$K(C_W) = \left\{g: g(A) = \int_A f(x)\mu(dx),\ f \in L^2(\mathcal{X},\mathcal{B},\mu)\right\},$$

and that the scalar product on $K(C_W)$ is given by $\langle g,g'\rangle_{K(C_W)} = \int f(x)f'(x)\mu(dx)$. Using the isometry $V:L^2(\mathcal{X},\mathcal{B},\mu) \to K(C_W)$, $Vf=g$, we can define the stochastic integral $I^W(f) = \pi(Vf)$ as in Theorem 2.3.1. Since the space $K^{\odot k}(C_W)$ is isomorphic to $L_S^2(\mathcal{X}^k,\mathcal{B}^{\otimes k},\mu^{\otimes k})$, the space of symmetric functions in $L^2(\mathcal{X}^k,\mathcal{B}^{\otimes k},\mu^{\otimes k})$, we can define the multiple Wiener integral $I_k^W(h_k)$ for $h_k \in L_S^2$. We shall prove that

$$Y_n(h) \Rightarrow \sum_{k=0}^{\infty} \frac{I_k^W(h_k)}{k!}$$

where $\Rightarrow$ denotes *convergence in distribution*. In particular, if $h_1 \in L_0^2(\mathcal{X},\mathcal{B},\mu)$ and

$$h(X_1,...,X_n) = \sum_{i=1}^{\infty} \sigma_1^n(h_1)$$

then

$$Y_n(h) = n^{-\frac{1}{2}}\sum_{i=1}^{n} h_1(X_i) \Rightarrow I_1^W(h_1)\ G\left(0,\|h_1\|_{L^2(\mathcal{X})}^2\right).$$

For $\varphi \in L_0^2(\mathcal{X},\mathcal{B},\mu)$, let us consider

$$\begin{aligned}
Y_n(h^\varphi) &= \sum_{k=0}^{\infty} n^{-\frac{k}{2}} \sum_{1\le s_1<...<s_k\le n} \varphi(X_{s_1})...\varphi(X_{s_k})\\
&= \sum_{k=0}^{\infty} \sum_{1\le s_1<...<s_k\le n} \frac{\varphi(X_{s_1})}{\sqrt{n}}...\frac{\varphi(X_{s_k})}{\sqrt{n}}\\
&= \prod_{j=1}^{n}\left(1+\frac{\varphi(X_j)}{\sqrt{n}}\right).
\end{aligned}$$

Note that with $n \to \infty$

$$\max_{1 \leq j} \leq n \frac{\varphi(X_j)}{\sqrt{n}} \to 0. \tag{3.28}$$

This allows us to take $\ln Y_n(h^\varphi)$ for n large enough. Using the Maclaurin expansion for $\ln(1+x)$, we obtain

$$
\begin{aligned}
\ln Y_n(h^\varphi) &= \sum_{j=1}^{n} \ln \left(1 + \frac{\varphi(X_j)}{\sqrt{n}} \right) \\
&= \sum_{j=1}^{n} \frac{\varphi(X_j)}{\sqrt{n}} - \frac{1}{2} \frac{\varphi^2(X_j)}{n} + \varepsilon_n(\varphi),
\end{aligned}
$$

where $\varepsilon_n(\varphi) \to$ by the WLLN and (3.28). Using the CLT and Slutsky's theorem, we conclude that

$$Y_n(h^\varphi) \Rightarrow e^{I_1^W(\varphi) - \frac{1}{2}\|\varphi\|_{L^2(\mathcal{X})}^2}.$$

By using the Cramèr-Wald device, it follows that for any finite subset $\{\varphi_1, ... \varphi_m\} \subseteq L_0^2(\mathcal{X}, \mathcal{B}, \mu)$,

$$(Y_n(h^{\varphi_1}), ..., Y_n(h^{\varphi_m})) \Rightarrow \left(e^{I_1^W(\varphi_1) - \frac{1}{2}\|\varphi_1\|_{L^2(\mathcal{X})}^2}, ..., e^{I_1^W(\varphi_m) - \frac{1}{2}\|\varphi_m\|_{L^2(\mathcal{X})}^2} \right).$$

Now, since we know the expansion of $e^{I_1(\varphi) - \frac{1}{2}\|\varphi\|_{L^2(\mathcal{X})}^2}$, we have

$$Y_n(h^\varphi) \Rightarrow \sum_{k=0}^{\infty} \frac{I_k^W(h_k^\varphi)}{k!}.$$

Let us take h with the defining sequence $h_k = \sum_{j=1}^{m} c_j h_k^{\varphi_j}(x_1, ..., x_k)$, where c_j are real numbers. Then

$$\sigma_k^n(h_k) = \sum_{1 \leq s_1 < ... < s_k \leq n} \sum_{j=1}^{m} c_j h^{\varphi_j}(x_1, ..., x_k) = \sum_{j=1}^{m} c_j \sigma_k^n \left(h_k^{\varphi_j} \right),$$

and for $h = (h_0, h_1, ..., h_k, ...)$,

$$Y_n(h) = \sum_{k=0}^{\infty} n^{-\frac{k}{2}} \sigma_k^n(h_k) = \sum_{j=1}^{m} c_j \sum_{k=0}^{\infty} n^{-\frac{k}{2}} \sigma_k^n \left(h_k^{\varphi_j} \right).$$

By the joint convergence above,

$$Y_n \Rightarrow \sum_{j=1}^{m} c_j \sum_{k=0}^{\infty} \frac{1}{k!} I_k^W \left(h_k^{\varphi_j} \right) = \sum_{k=0}^{\infty} \frac{1}{k!} I_k^W(h_k),$$

as I_k^W is linear. Now using the following identity on pg. 744 in [22],

$$E\left(\sigma_k^n\left(h_k - h_k'\right)\sigma_l^n\left(h_l - h_l'\right)\right) = \begin{cases} \binom{n}{k}\left\|h_k - h_k'\right\|_{L^2(\mathcal{X})}^2 & \text{if } l = k \\ 0 & \text{if } l \neq k \end{cases}$$

we arrive at

$$E\left(Y_n(h) - Y_n(h')\right)^2 = \sum_{k=0}^{\infty}\binom{n}{k}n^{-k}\left\|h_h - h_k'\right\|_{L^2(\mathcal{X})}^2 \leq \left\|h_h - h_k'\right\|_{\mathcal{H}}^2$$

We now state the main result.

Theorem 3.5.1. *For any $h = (h_0, h_1, ..., h_k, ...) \in \mathcal{H}$,*

$$Y_n(h) \Rightarrow W(h) = \sum_{k=0}^{\infty}\frac{I_k(h_k)}{k!}.$$

Proof. Let $h \in \mathcal{H}$ and $\varepsilon > 0$. Choose $h' = \sum_{j \in J}c_j h^{\varphi_j}$, where J is a finite set, so that $\|h - h'\|_{\mathcal{H}} < \varepsilon/2$. For $t \in \mathbb{R}$,

$$\left|E\left(e^{itY_n(h)} - e^{itW(h)}\right)\right| \leq \left|E\left(e^{itY_n(h)} - e^{itY_n(h')}\right)\right| + \left|E\left(e^{itY_n(h')} - e^{itW(h')}\right)\right|$$
$$+ \left|E\left(e^{itW(h')} - e^{itW(h)}\right)\right|.$$

Using the inequality $|e^x - 1| \leq |x|$ and the Schwartz inequality we can dominate the first and third term by $|t|\|h - h'\|_{L^2(\mathcal{X})}$. As $n \to \infty$, the middle term converges to zero for each t. Thus, the result follows. ☐

Remark 3.5.1. *In conclusion of this chapter we note the following:*

(1) Instead of using the CLT if we use the Donsker invariance principle and construct a Gaussian random field $\{W(s,A), s \leq T, A \in \mathcal{B}\}$ with covariance $C((t,A),(s,B)) = (t \wedge s)\mu(A \cap B)$, then one can prove an invariance principle for

$$Y_n^t(h) = \sum_{k=0}^{\infty}[nt]^{-\frac{k}{2}}\sum_{1 \leq s_1 < ... < s_k \leq [nt]}\sigma_k^n(h_k),$$

giving the result of Mandelbaum and Taqqu [78]. For details see [82].

(2) The classical result on U-statistics due to Hoeffding [44] and Serfling [114] can be easily derived from Theorem 3.5.1 (see [22]).

Chapter 4

Skorokhod and Malliavin Derivatives for Gaussian Random Fields

In Chapter 3, we defined the Skorokhod derivative as a map from $L^2(\Omega, \mathcal{F}, P)$ to $L^2(\Omega, K)$ whose adjoint operator is the Skorokhod integral. In [76], Malliavin defined the concept of a derivative for the case of real-valued Brownian motion with the covariance $C(t,s) = t \wedge s$, $t, s \geq 0$. Following the work in [94], we show that the two concepts of a derivative coincide. As the Malliavin derivative led to a development of calculus, named after Malliavin, we shall present an introduction to this area for general Gaussian random fields.

4.1 Malliavin Derivative

Let us introduce the ideas of Malliavin in [76] as presented by Nualart in [94] and [95]. Denote by $C_p^\infty(\mathbb{R}^n)$ the subset of $C^\infty(\mathbb{R}^n)$ consisting of functions $f : \mathbb{R}^n \to \mathbb{R}^n$, such that f and all its derivatives have polynomial growth.

Definition 4.1.1. *A smooth functional is a random variable $F : \Omega \to \mathbb{R}$ of the form*

$$F = f\left(W_{t_1}, \ldots, W_{t_n}\right), \tag{4.1}$$

where $f \in C_p^\infty(\mathbb{R}^n)$, $0 \leq t_1, \ldots, t_n \leq 1$. The class of smooth functionals will be denoted by S.

We now define the Malliavin derivative of a smooth functional.

Definition 4.1.2. *The Malliavin derivative of a smooth functional F given by 4.1 is defined as*

$$\left(D_t^M F\right) = \sum_{i=1}^n \partial_i f\left(W_{t_1}, \ldots, W_{t_n}\right) I_{[0,t_i]}(t), \, t \in [0,1].$$

Recall the identification of the stochastic integral $\int_0^t g(t)\, dW_t = \pi \circ V(g)$ from Theorem 2.3.1 and Example 3.3.1. The following result, Proposition 3.3 in [94], together with Lemma 3.3.1, shows that Malliavin and Skorokhod derivatives for Brownian motion are the same.

Proposition 4.1.1. *Suppose that $F = \sum_{m=0}^\infty I_m(f_m)$ is a square integrable random variable. Then $F \in D^{1,2}$ if and only if*

$$\sum_{m=1}^\infty mm! \, \|f_m\|_{L^2([0,1]^n)}^2 < \infty, \tag{4.2}$$

and in this case

$$\left(D_t^M F\right) = \sum_{m=1}^{\infty} m I_{m-1}\left(f_m(\cdot,t)\right),$$

and $E \int_0^1 \left(D_t^M F\right)^2 dt$ is given by the series in 4.2.

The following properties of Malliavin (Skorokhod) derivative are stated, respectively, in [95] as Proposition 1.2.3 and Proposition 1.2.1.

Proposition 4.1.2 (Chain Rule)**.** *Let $F \in D^{1,2}$ and $g : \mathbb{R} \to \mathbb{R}$ be continuously differentiable with bounded derivative. Then $g(F) \in D^{1,2}$ and*

$$D^M g(F) = g'(F) D^M F.$$

Proposition 4.1.3. *The Malliavin (Skorokhod) derivative $D^M = D$ is closable from $L^2(\Omega) \to L^2(\Omega, K(C))$.*

As a consequence, for the closure of the Malliavin (Skorokhod) derivative we have the following theorem.

Theorem 4.1.1. *Let $F \in L^2(\Omega)$ and $F_k \in D^{1,2}$ be such that $F_k \to F$ in $L^2(\Omega)$ and $D^M F_k \to \eta$ in $L^2(\Omega, K(C))$, then $F \in D^{1,2}$ and $\eta = D^M F$.*

4.2   Duality of the Skorokhod Integral and Derivative

In this section we establish a result that the Skorokhod integral is an adjoint operator to the Skorokhod derivative. We begin with the following lemma.

Lemma 4.2.1. *Let $u \in L^2(\Omega, K)$, then u is Skorokhod integrable if and only if there exists a constant $M > 0$ such that*

$$|E\langle u., D.F\rangle_K| \le M\|F\|_{L^2(\Omega)}, \tag{4.3}$$

for any $F \in L^2(\Omega, \mathcal{F}, P)$ which is Skorokhod differentiable. In this case,

$$|E\langle u., D.F\rangle_K| = E\left(I^s(u)F\right). \tag{4.4}$$

In particular, I^s is an adjoint operator to D. on $L^2(\Omega, \mathcal{F}, P)$.

Proof. It suffices to prove the result for $F = I_p(g)$, $g \in K^{\odot p}$ and $u_t = \sum_{p=0}^{\infty} I_p\left(f_p(\cdot,t)\right)$, with $f_p \in K^{\otimes p+1}$ and $f_p(\cdot,t) \in K^{\odot p}$ for any fixed t. Note that $D_t F = p I_{p-1}(g(\cdot,t))$. Assume that $\{e_j, j \in J\}$ is an orthonormal basis in K. By the Parseval identity,

$$\left\langle I_q\left(f_q(\cdot,*)\right), p I_{p-1}(g(\cdot,*))\right\rangle_K$$

$$= \sum_{j=1}^{\infty} \left\langle I_q\left(f_q(\cdot,*)\right), e_j(*)\right\rangle_K \left\langle p I_{p-1}(g(\cdot,*)), e_j(*)\right\rangle_K = \sum_{j=1}^{\infty} Z_{q,p}^j.$$

Hence, by the orthogonality of I_q and I_p for $q \neq p$,

$$
\begin{aligned}
E \langle u., D.F \rangle_K &= \sum_{j,q} E Z_{q,p}^{j} \\
&= p(p-1)! \langle f_{p-1}, g_{p-1} \rangle_{K \otimes p} \\
&= p! \langle \tilde{f}_{p-1}, g_{p-1} \rangle_{K \otimes p} \\
&= E \left(I^s(u) F \right).
\end{aligned}
$$

In the calculations above, f could be replaced with $\tilde{f}$ due to the symmetry of g. Thus, it follows that Skorokhod integrability of u implies (4.4) and (4.3).

Conversely, if $u \in L^2(\Omega, K)$ then using (4.3), we have for any positive integer n,

$$
\left| E \left(\sum_{p=1}^{n} I_{p-1} (\tilde{f}_p) F \right) \right| \leq M \| F \|_{L^2(\Omega)},
$$

giving

$$
\left| E \left(\sum_{p=1}^{n} I_{p-1} (\tilde{f}_p) \right) \right|^2 \leq M.
$$

In conclusion,

$$
\sum_{p=1}^{\infty} p! \left\| \tilde{f}_{p-1} \right\|_{K \otimes p}^2 < \infty,
$$

proving Skorokhod integrability of u. $\qquad \square$

4.3 Duration in Stochastic Setting

Let $P(t,T)$ be the price at time t of a zero coupon bond, which pays \$1 at maturity time T. Suppose that bond prices are modeled by non-negative adapted processes $\{P(t,T), 0 \leq t \leq T\}$ for each $t > 0$ on a filtered probability space $(\Omega, \mathcal{F}, (\mathcal{F}_t)_{t \geq 0}, P)$. We assume that the bond prices $P(t,T)$ are described by Heath-Jarrow-Morton model (HJM model) [43]

$$
P(t,T) = \exp \left(-\int_0^T f(t,s) \, ds \right),
$$

where $f(t,x)$, $0 \leq t \leq s < \infty$ are instantaneous forward rates modeled by SDE

$$
df(t,T) = \alpha(t,T) \, dt + \sigma(t,T) \, dW(t), \quad 0 \leq t \leq T < \infty.
$$

One can re-parameterize the forward rates by time to maturity $x = T - t$ getting forward curves $f(t,x) = f(t,t+x)$. Setting $\alpha_t(x) = \alpha(t,t+x)$, $\sigma_t(x) = \sigma(t,t+x)$, one gets the so-called Musiela equation

$$
df_t(x) = \frac{d}{dx} f_t(x) + \alpha_t(x) \, dt + \sigma_t(x) \, dW_t. \tag{4.5}
$$

As equation (4.5) does not capture the maturity-specific risk, a more realistic model is presented in [57].

$$df_t(x) = \frac{d}{dx} f_t(x) + \alpha_t(x)\, dt + \sigma_t(x)\, dW_t(x),$$

where $W_t(x)$ stands for risk arising from the time to maturity x. Hence $W_t(x)$ is considered as Brownian sheet. They recast the equation above as

$$df_t(x) = \frac{d}{dx} f_t(x) + \alpha_t(x)\, dt + \sum_{k \geq 1} \sigma_t^{(k)}(x)\, dW_t^{(k)}(x), \qquad (4.6)$$

with $\{\sigma^{(k)}(\cdot),\, k \geq 1\}$ deterministic measurable functions and $\{W^{(k)}(t),\, k \geq 1\}$ independent univariate Brownian motions.

We shall assume that forward curves are modeled by functions in a Hilbert space H such that $x \to f(x)$ is continuous. Also, $(S_t f)(x) = f(t + x)$ is a strongly continuous semigroup on H. An example of such a Hilbert space H is given in [28].

Let Q be a symmetric non-negative trace class operator on a Hilbert space U and $U_0 = Q^{1/2} U$ with norm $\|h\|_0 = \|Q^{-1/2}(h)\|_U$, $h \in U_0$. Denote by $\mathcal{L}_2(U, H)$ the space of Hilbert–Schmidt operators from U to H with the operator norm $\|\cdot\|_{\mathcal{L}_2}$. Let $\{u_k, k \geq 1\}$ be on basis in U and suppose there exist a Borel measurable map $\sigma : [0,T] \to \mathcal{L}(U_0, H)$ such that $\sigma_t\left(Q^{1/2}(u_k)\right) = \sigma_t^{(k)}(\cdot)$ and $\sigma_t \circ Q^{1/2} \in \mathcal{L}_2(U, H)$ for all (t,k) in (4.6). In addition denote, $A = \frac{d}{dx}$. Viewing the collection $\left\{W_t^{(k)},\, 0 \leq t \leq T\right\}$ as a cylindrical Wiener process, we can rewrite Equation (4.6),

$$df_t = A f_t + \alpha_t\, dt + \sigma_t\, dW_t, \qquad (4.7)$$

see Section 2.2.4 in [36] for the definition of the stochastic integral in this case. Under appropriate conditions (see [49]) there exists a unique solution to Equation (4.7).

Assume σ_t is invertible a.e. for $0 \leq t \leq T$, and

$$\sup_{t \in [0,T]} E\left(\delta \left\|\sigma_t^{-1}(A f_t + \alpha_t)\right\|_0^2\right) < \infty.$$

Then applying the Girsanov Theorem (cf. Chapter 6, see also [5]) to (4.7) gives

$$df_t = \sigma_t\, d\hat{W}_t,$$

where $\hat{W}_t = W_t - \int_0^t \psi(s)\, ds$ with $\psi(t) = \sigma_t^{-1}(A f_t + \alpha_t)$ and $\hat{W}_t$ is a Q-Brownian motion under the probability measure

$$\hat{P}(A) = E\left(1_A \exp\left(\int_0^T \langle \psi(s), dW(s)\rangle_0 - \frac{1}{2}\int_0^T \|\psi(s)\|_0^2\, ds\right)\right).$$

Consequently, f_t is a Gaussian martingale with respect to $\hat{P}$. Define

$$\hat{f}_t = f_t - f_0 = \int_0^t \sigma_s d\hat{W}_s.$$

Then $\hat{f}_t(x)$ is a centered Gaussian random field with respect to time and time to maturity under $\hat{P}$.

Now we use these forward curves to define the concept of duration which serves as a tool to measure interest rate sensitivities of bond portfolios with respect to the whole yield surface $(t,x) \to f_t(x)$.

Let us denote for $u = (t,x)$ the Gaussian random field $\hat{f}(u) = \hat{f}_t(x)$ with covariance $C(u,r) = E\hat{f}(u)\hat{f}(r)$. Denote by $K(C)$ the RKHS of $\hat{f}$. Using continuity of evaluation functionals on H and Banach-Steinhaus theorem, we can see that K is isometrically isomorphic to the space

$$H(\hat{f}) = \left\{ \lambda : [0,T] \to H^*, \text{ Borel measurable, and } \int_0^T \|\lambda_s \circ \sigma_s\|_{\mathcal{L}_2^0}^2 ds < \infty \right\},$$

where $\|B\|_{\mathcal{L}_2^0} = \|B \circ Q^{1/2}\|_{L^2(U,H)}$ for $B \in \mathcal{L}(H,H)$. Here H is as given before with H^* its dual.

Recall that for $F \in L^2(\Omega,\hat{P})$, we denote by $D_u F$ its Mallianvin (Skorokhod) derivative, and $D^{1,2} \subseteq L^2(\Omega,\hat{P})$ is its domain.

In view of the financial applications to be looked at, we can regard the derivative as a sensitivity measure with respect to the fluctuations of the yield surface $(t,x) \to f_t(x)$.

Considering $\mathcal{X}$ as the support of the image measure μ of $\hat{f}$ under $\hat{P}$ in $C([0,T],H)$ we know by[7] that $\mathcal{X} = \overline{K}$ (closure) in $C([0,T],H)$ and as shown earlier $F \in L^2(\mu,K)$. If for $F \in L^2(\mu,K)$, $(F(x+\varepsilon k) - F(x))/\varepsilon$ converges in $L^2(\mu,K)$ (Gateaux derivative) for $k \in K$, then $D.F \in L^2(\mu,K)$ exists and the above limit equals $\langle D.F, k \rangle_K$ (see [31]). Since P is equivalent to $\hat{P}$ the convergence above holds in probability with respect to the image measure of forward curves under original P. If $F = \xi_T$ is the terminal value of a bond portfolio, one interprets $D.F$ as a sensitivity measure of the fluctuations of the whole yield surface in this portfolio. Using this idea, one can define an expanded duration as in [57].

Definition 4.3.1 (Stochastic Duration). *Let F be a square integrable functional of the forward curve $\hat{f}$ with respect to $\hat{P}$. Assume that F is Malliavin differentiable with respect to $\hat{f}$, that is, $F \in D^{1,2}$. Then the stochastic duration of F is the process $D.F \in L^2(\Omega,\hat{P},K)$.*

We present three examples of stochastic duration.

Example 4.3.1 (Zero Coupon Bond). *Let $P(t,T)$ be the price at time t of a zero coupon bond, which pays $\$1$ at maturity time T. Then using instantaneous* *forward rates $f(t,s)$, $0 \le t \le s$, we have*

$$
\begin{aligned}
P(t,T) &= \exp\left(-\int_t^T f(t,s)\,ds\right) \\
&= \exp\left(-\int_0^{T-t} f_t(x)\,dx\right).
\end{aligned}
$$

Observe that

$$
\begin{aligned}
D_{r,y}\left(\int_0^{T-t} f_t(x)\,dx\right) &= \int_0^{T-t} D_{r,y}(f_t(x))\,dx \\
&= \int_0^{T-t} 1_{[0,t]}(r)\,dx \\
&= (T-t)1_{[0,t]}(r).
\end{aligned}
$$

The chain rule, Lemma 1.2.3 in [95], shows that the stochastic duration $D.P(t,T)$ in the HJM model is given by

$$
D_{r,y}P(t,T) = \begin{cases} -(T-t)P(t,T), & 0 \le r \le t; \\ 0, & otherwise. \end{cases}
$$

Remark 4.3.1. *Note that $D_{r,y}P(t,T)/P(t,T)$, $0 \le r \le t$ has the form of the classical duration. The latter expression seems to suggest that one should use $D.F/F$ as the generalized duration rather than $D.F$. However, a general interest rate claims F may be zero with positive probability. Therefore, it is reasonable to use $D.F$ as the expanded concept of duration.*

The classical duration introduced by Macauley in [75] presumes yield surfaces are flat or piecewise flat. Such a model is fundamentally different from a stochastic interest model.

Example 4.3.2 (Interest Rate Cap). *Consider a cap of the form*

$$
F = (R(t,T) - K)^+,
$$

where K is the cap rate and $R(t,T)$ is given by

$$
R(t,T) = \frac{1}{T-t}\int_t^T r(s)\,ds
$$

is the average interest rate based on overnight interest rate(start rate) $r(t) = f(t,t)$. We observe that

$$
\begin{aligned}
D_{r,y}\left(\frac{1}{T-t}\int_t^T r(s)\,ds\right) &= \frac{1}{T-t}\int_t^T D_{r,y}(r(s))\,ds \\
&= \frac{1}{T-t}\int_t^T D_{r,y}(f_s(0))\,ds \\
&= 1_{[0,t]}(r).
\end{aligned}
$$

We approximate $\varphi(x) = (x - K)^+$ by φ_n with

$$\varphi_n(x) = \varphi(x), \quad |x - K| \geq \frac{1}{n}$$

and $0 \leq \varphi'_n \leq 1$ for all x. Then it follows from the chain rule and closability of the derivative in Proposition 1.2.1 in [95], that

$$D_{r,y}F = 1_{[K,\infty]}(R(t,T))1_{[0,t]}(r).$$

Example 4.3.3 (Asian Option). *Consider an Asian type of option defined as*

$$F = \frac{1}{(\bar{x}_2 - \bar{x}_1)(T_2 - T_1)} \int_{\bar{x}_1}^{\bar{x}_2} \int_{T_1}^{T_2} f_t(x)\,dt\,dx.$$

Then

$$\begin{aligned} D_{r,y}F &= \frac{1}{(\bar{x}_2 - \bar{x}_1)(T_2 - T_1)} \int_{\bar{x}_1}^{\bar{x}_2} \int_{T_1}^{T_2} 1_{[0,t]}(r)\,dt\,dx \\ &= 1_{[0,t]}(r). \end{aligned}$$

4.4   Special Structure of Covariance and Ito Formula

Let us recall Exercise 2.3.4(b). As we are interested in covariance properties, we shall assume that we are given a Gaussian process of the form $\{X(A), A \in \mathcal{B}(\mathcal{X})\}$ where $(\mathcal{X}, \mathcal{B}(\mathcal{X}))$ is a measurable space. Further we assume that

$$X(A) = P_{L(X)}\xi(A), \quad A \in \mathcal{B}(\mathcal{X}), \tag{4.8}$$

where $\{\xi(A), A \in \mathcal{B}(\mathcal{X})\}$ is independently scattered Gaussian measure (that is, there exists a non-negative measure v such that $E\xi(A)\xi(B) = v(A \cap B)$, $A, B \in \mathcal{B}(\mathcal{X})$) and $L(X) = \overline{\text{span}}\{X(A), A \in \mathcal{B}(\mathcal{X})\}$.

Exercise 4.4.1. *Prove that the RKHS of the covariance $v(A \cap B)$ is of the form $\{\mu(A), A \in \mathcal{B}(\mathcal{X})\}$ with*

$$\mu(A) = \int_A f_\mu(x)\,v(dx), \quad f_\mu \in L^2(\mathcal{X}, v).$$

Now let us examine the form of the RKHS of the covariance of X,

$$C_X(A,B) = EX(A)X(B) \quad \text{for } A, B \in \mathcal{B}(\mathcal{X})$$

We note that from the general theory $\mu \in K(C_X)$ implies that for $A \in \mathcal{B}(\mathcal{X})$

$$\mu(A) = \langle \mu, C_X(\cdot, A)\rangle_{K(C_X)}.$$

Hence we can see that μ is a signed measure with

$$|\mu(A)| \leq \|\mu\|_{K(C_X)} C_X^{1/2}(A,A)$$

and $C_X(A,A) \le v(A)$, giving the fact that $\mu \ll v$. We can therefore write

$$\mu(A) = \int_A f_\mu(x)\, v(dx)$$

and if $f_{\mu_1}, f_{\mu_2} \in L^2(\mathcal{X}, v)$, then we define

$$\langle \mu_1, \mu_2 \rangle_{K(C_X)} = \int_{\mathcal{X}} f_{\mu_1}(x) f_{\mu_2}(x)\, v(dx). \qquad (4.9)$$

Thus, $K(C_X)$ is isometric to the completion of $L^2(\mathcal{X}, v)$ in the norm induced by the scalar product in (4.9). We can therefore extend the definition of the stochastic integral

$$\int f_\mu(u)\, dX(u) \qquad (4.10)$$

to this completion.

We now consider the special case when $\mathcal{X} = R_+$ and $\mathcal{B}(\mathcal{X})$ are the Borel subsets of R_+. Let $\{X_t, t \in [0,T]\}$ be a càdlàg stochastic process defined on a probability space $(\Omega, \mathcal{F}, P)$ and satisfying condition (4.8). Assume that the filtration $\mathcal{F}_t = \sigma\{X_s, s \le t\}$ satisfies the *usual conditions* and is *free of times of discontinuity* as in [39].

Define the partition

$$\Delta(n,i,X) = X_{T(i+1)/2^n} - X_{Ti/2^n}$$

and

$$S(n,X,X) = EX_0^2 + \sum_{i=1}^{\infty} E(\Delta(n,i,X))^2.$$

We recall the definition of a process of *finite energy* in the sense of [39].

Definition 4.4.1. *A stochastic process X has finite energy if the set of numbers $S(n,X,X)$ is bounded.*

Using the fact that

$$E(\Delta(n,i,X))^2 \le E(\Delta(n,i,\xi))^2,$$

we conclude that X has finite energy. From Theorem 1 of [39] we obtain that X has a decomposition

$$X = M + A, \qquad (4.11)$$

where M is a martingale and A is previsible and A has the following property: there exists a subsequence n_j such that for every square integrable martingale g

$$\lim_j \left\{ E(X_0 g_0) + \sum_i E\left(\Delta(n_j, i, X)\Delta(n_j, i, g)\right) \right\} = 0.$$

In addition, if $N + B$ is another such decomposition, then $A - B$ is a continuous martingale.

We also know from Corollary 2 in [39] that if $\limsup_n S(n, A) = 0$, then the decomposition (4.11) is unique.

Let us now assume the $\{X_t, t \le T\}$ is cádlág and $\sup_n S(n, X, X) < \infty$, that is, X has finite energy. Then, clearly, we can write $X_t = M_t + A_t$ where M_t is a square integrable martingale and A_t is as above. If we further assume that $\{A_t, 0 \le t \le T\}$ has zero *quadratic variation*, that is,

$$\limsup_n S(n, A, A) = 0,$$

then we can get the Itô Formula (Theorem 6 in [39]). Let $X = M + A$ be continuous and A have zero quadratic variation. Let f be continuously differentiable with bounded derivative. Then $Y = f(X)$ will be a sum of square integrable martingale and a process with zero quadratic variation and the martingale part of Y is

$$Y_0 + \int_0^t f'(X_s)\, dM_s.$$

Remark 4.4.1. (a) *In view of [80], we know that Gaussian processes with covariance of bounded variation in two dimensions has the property (4.8) but the converse is not true. The results presented so far were motivated by the paper [64], which we shall study next.*

(b) *Using the results in Chapters 2 and 3 on chaos expansion and Sections 4.1.2 and 4.2, we can define the Skorokhod integral and the Malliavin derivative in case of processes having the form (4.8) by using integral $\int f_\mu(u)\, dX(u)$ defined in (4.10). This generalizes the work in [64].*

We now assume that the covariance function $C(A, B)$, $A, B \in \mathcal{B}(\mathcal{X})$ of a Gaussian random field $\{X(A), A \in \mathcal{B}(\mathcal{X})\}$ is a finite measure μ on $\mathcal{B}(\mathcal{X}) \otimes \mathcal{B}(\mathcal{X})$ in the sense that for partitions $\{A_i^n \times B_j^n, i, j = 1, 2, ..., n\}$, $n = 1, 2, ...,$ of $\mathcal{X} \otimes \mathcal{X}$, $\lim_n \sum_{i,j=1}^n C(A_i^n, B_j^n)$ exists as partitions become finer.

Using the fact that for $A_1 \cap A_2 = \varnothing$, $E(X(A_1 \cup A_2) - X(A_1) - X(A_2))^2 = 0$ and $E(X(A_n))^2 \downarrow 0$ if $A_n \downarrow \varnothing$ we can see that X is an $L_2(\Omega)$–valued measure. Let $\Lambda(\mu)$ denote the space of functions f, such that $f : \mathcal{X} \to R$ with $\int_{\mathcal{X}} \int_{\mathcal{X}} f(u_1) f(u_2)\, \mu(du_1, du_2) < \infty$. We can define for $f \in \Lambda(\mu)$ the integral $\int f(u)\, dX(u)$ and can extend it to all f in the completion of $\Lambda(\mu)$ to a Hilbert space. As stated in Remark 4.4.1, we can extend this definition to obtain chaos expansion.

Now we consider the case $\mathcal{X} = [0, T]$ and $EX(t)X(s) = C(t, s)$. Define for $I = (a_1, b_1] \times (a_2, b_2]$ by

$$\Delta_I C = C(b_1, b_2) + C(a_1, a_2) - C(a_1, b_2) - C(b_1, a_2)$$

and assume that

$$\sup_{\tau} \sum_{i,j=1}^{n} \left| \Delta_{(t_i,t_{i+1}] \times (t_j,t_{j+1}]} C \right| < \infty,$$

where $\tau = \{0 = t_0 < \cdots < t_n = T\}$ is a partition of $[0,T]$ and under our assumption $\Delta_t C \geq 0$. Suppose Π_n are finer and finer partitions and consider

$$\sum_{i=1}^{n} E\left(X(t_{i+1}^n \wedge t) - X(t_i^n \wedge t)\right)^2 = \mu\left(\bigcup_{i=0}^{n} \{[t_i^n, t_{i+1}^n] \times [t_i^n, t_{i+1}^n]\} \bigcap ([0,t] \times [0,t])\right).$$

Then by the fact that the RHS is increasing as the partitions get finer, we obtain that the following limit exists:

$$\lim_{\Pi_n} E\left(\sum_{i=1}^{n} \left(X(t_{i+1}^n \wedge t) - X(t_i^n \wedge t)\right)^2\right)$$

Hence the limit

$$\lim_{\Pi_n} \sum_{i=1}^{n} \left(X(t_{i+1}^n \wedge t) - X(t_i^n \wedge t)\right)^2$$

exists in probability giving the existence of the quadratic variation of $\{X_t, t \in [0,T]\}$. Using the method due to Föllmer (see [29], for details see [119], p. 21) we can obtain the following Itô Formula.

Theorem 4.4.1. *Let $X : [0,\infty) \to R^1$ be a continuous Gaussian process with covariance of bounded variation, with quadratic variation $\langle X \rangle_t$ and let $f \in C^2(R^1)$ (twice continuously differentiable function). Then*

$$f(X(t)) = f(X(0)) + \int_0^t f'(X(s)) \, dX(s) + \frac{1}{2} \int_0^t f''(X(s)) \, d\langle X \rangle_s,$$

where

$$\int_0^t f'(X(s)) \, dX(s) = \lim_{n} \sum_{t_i \in \Pi_n, t_i \leq t} f'(X(t_i^n)) \left(X(t_{i+1}^n) - X(t_i^n)\right)$$

For interesting examples of Gaussian processes with covariance of bounded variation we refer the reader to [64], where the proofs are given for the following cases.

(1) Gaussian Martingale.

(2) Fractional Brownian Motion with the Hurst parameter $H > \frac{1}{2}$.

(3) Bifractional Brownian Motion $H \in (0,1)$, $K \in (0,1)$, $2HK \geq 1$.

(4) Gaussian process with stationary increments with

$$Q(t-s) = E\left((X(t) - X(s))^2\right) \quad \text{and} \quad Q(t) = E\left((X(t))^2\right).$$

If Q'' is a Radon measure, then X has covariance of bounded variation and $Q'(0+)$ exists if and only if X has finite energy.

Chapter 5

Filtering with General Gaussian Noise

In this chapter we restrict ourselves to index sets which are subsets of $\mathbb{R}$. Because of the natural ordering of an index set we are able to introduce the filtering problem and the Itô formula. We begin by describing the *filtering problem*.

5.1 Bayes Formula

The *signal* or *system process* $\{X_t, 0 \leq t \leq T\}$ is unobservable. Information about X_t is obtained by observing another process $\{Y_t, 0 \leq t \leq T\}$, which is a function of X corrupted by *noise N*, that is,

$$Y_t = \beta_t + N_t, \quad 0 \leq t \leq T,$$

where β_t is adapted to the filtration generated by the signal, $\mathcal{F}_t^X = \sigma\{X_s, s \leq t\}$, augmented by inclusion of null sets, and N_t is a noise process. The observation σ-field $\mathcal{F}_t^Y = \sigma\{Y_s, 0 \leq s \leq t\}$ contains all available information about the signal process X_t. The primary aim of filtering theory is to obtain an estimate of X_t based on the information $\mathcal{F}_t^Y$. This estimator is given by the conditional distribution of X_t given Y_t, or equivalently, the conditional expectations $E\left(f(X_t) \big| \mathcal{F}_t^Y\right)$ for a rich enough class of functions f. Since this estimator minimizes the squared error loss, it is called the optimal filter. We now state precise assumptions and conditions for obtaining the optimal filter.

Suppose $\{X_t, 0 \leq t \leq T\}$ is a real valued signal process and the observation process is given by

$$Y_t = \beta(t, X) + N_t, \quad 0 \leq t \leq T. \tag{5.1}$$

Here $\beta : [0, T] \times \mathbb{R}^{[0,T]} \to \mathbb{R}$ is a *non-anticipative function*, that is, $\beta(t, x)$ is $\mathcal{C}_t$-measurable, where $\mathcal{C}_t$ is the cylindrical σ-field generated by the cylinder sets in $\mathbb{R}^{[0,T]}$ with bases in $[0, t]$. The noise process $\{N_t, 0 \leq t \leq T\}$ is assumed independent of the signal process $\{X_t, 0 \leq t \leq T\}$.

We need to find a Bayes formula for $E\left(f(X_t) \big| \mathcal{F}_t^Y\right)$. To explain this idea, let us consider the simplest case.

Let X, Y, Z, be random variables defined on a probability space $(\Omega, \mathcal{F}, P)$ and assume that Z is standard normal independent of X, and $Y = X + Z$. Consider the problem of computing $E(f(X)|Y)$. Suppose P and Q are probability measures on Ω with $P \ll Q$ and let $\mathcal{G} \subseteq \mathcal{F}$ be a sub σ-field. Then,

$$E_P(f(x)|\mathcal{G}) = \frac{E_Q\left(f(X)\frac{dP}{dQ}\Big|\mathcal{G}\right)}{E_Q\left(\frac{dP}{dQ}\Big|\mathcal{G}\right)}. \tag{5.2}$$

If we define

$$dQ = e^{-XY+\frac{1}{2}X^2} dP$$

then Q is a probability measure.

Exercise 5.1.1. *Under the measure Q, Y is a standard normal random variable independent of X and the random variable X has the same distribution under P and Q.*

Thus, we obtain the Bayes formula,

$$E_P(f(X)|Y) = \frac{\int f(x)e^{xY-\frac{1}{2}x^2} dP_X(x)}{\int E^{xY-\frac{1}{2}x^2} dP_X(x)} \tag{5.3}$$

We now give the analogue of the above result for the general problem. Let $\{N_t, 0 \le t \le T\}$ be a centered Gaussian process, $EN_t = 0$, whose covariance $R(s,t) = EN_sN_t$ is continuous on the rectangle $[0,T] \times [0,T]$. Let $\{\xi_t, 0 \le t \le T\}$ be a process with values in a complete separable metric space $\mathcal{X}$, independent of $\{N_t, 0 \le t \le T\}$. Assume that

$$Y_t = f(t,\xi) + N_t, \quad 0 \le t \le T,$$

where f is a non-anticipative measurable function on $[0,T] \times \mathcal{X}^{[0,T]}$. Denote by $K(R,t)$ the RKHS of R restricted to the rectangle $[0,t] \times [0,t]$, with the norm and scalar product $\|\cdot\|_t$ and $\langle \cdot, \cdot \rangle_t$. In particular, $K(R) = K(R,T)$. Denote the stochastic integral with respect to the process $\{N_s, 0 \le s \le t\}$ by π_t,

$$\pi_t : K(R,t) \to \overline{\text{span}}^{L^2(\Omega)} \{N_s, 0 \le s \le t\}.$$

Then for $g,h \in K(R,t)$, $E\pi_t(g)\pi_t(h) = \langle g,h \rangle_t$.

We will need the following theorem.

Theorem 5.1.1. *Assume that $f(\cdot,\xi) \in K(R)$ a.s. Define for each $0 \le t \le T$,*

$$dQ_t = e^{-\pi_t(f)-\frac{1}{2}\|f\|_t^2} dP.$$

Then Q_t is a probability measure and under Q_t, we have

(i) *$\{Y_s, 0 \le s \le t\}$ is a centered Gaussian process with covariance R, and is independent of $\{\xi_s, 0 \le s \le t\}$.*

(ii) *$\{\xi_s, 0 \le s \le t\}$ has the same distribution as under the measure P.*

Proof. Let $0 \le t \le T$ be fixed. Since $f(\cdot,\xi) \in K(R)$ a.s. then $f(\cdot,\xi)|_{[0,t]} \in K(R,t)$ a.s. Also, $\pi_t(f)$ is normally distributed with mean zero and variance $\|f\|_t^2$. Hence Q_t is a probability measure since ξ and N are independent processes. Now let $0 \le s_1 \le ... \le s_m \le t, 0 \le t_1 \le ... \le t_n \le T$; $g_1,...,g_n : \mathcal{X} \to \mathbb{R}$ be measurable functions and $\alpha_1,...,\alpha_n; \gamma_1,...,\gamma_m$ be real numbers. Consider the joint characteristic function, using $Y_{s_k} = f(s_k,\xi) + N_{s_k}$,

$$E_{Q_t}\left(\exp\left(i\sum_{j=1}^{n}\alpha_j g_j(\xi_{t_j}) + i\sum_{k=1}^{m}\gamma_k Y_{s_k}\right)\right)$$

$$= E_P\left(\exp\left(i\sum_{j=1}^{n}\alpha_j g_j(\xi_{t_j}) + i\sum_{k=1}^{m}\gamma_k Y_{s_k}\right)\exp\left(-\pi_t(f) - \frac{1}{2}\|f\|_t^2\right)\right)$$

$$= E_P\left(\exp\left(i\sum_{j=1}^{n}\alpha_j g_j(\xi_{t_j}) - \frac{1}{2}\|f\|_t^2 + i\sum_{k=1}^{m}\gamma_k f(s_k,\xi) + i\sum_{k=1}^{m}\gamma_k N_{s_k} - \pi_t(f)\right)\right)$$

$$= E_P\left(\exp\left(i\sum_{j=1}^{n}\alpha_j g_j(\xi_{t_j}) - \frac{1}{2}\|f\|_t^2 + i\sum_{k=1}^{m}\gamma_k f(s_k,\xi)\right)\right.$$

$$\left. \times E_P\left(\exp\left(i\sum_{k=1}^{m}\gamma_k N_{s_k} - \pi_t(f)\right)\Big|\mathcal{F}_T\right)\right)$$

$$= E_P\left(\exp\left(i\sum_{j=1}^{n}\alpha_j g_j(\xi_{t_j}) - \frac{1}{2}\|f\|_t^2 + i\sum_{k=1}^{m}\gamma_k f(s_k,\xi)\right)\right.$$

$$\left. \times \exp\left(-\frac{1}{2}\sum_{k,l=1}^{m}\gamma_k\gamma_l R(s_k,s_l) - \frac{1}{2}2\sum_{k=1}^{m}\gamma_k f(s_k,\xi) + \frac{1}{2}\|f\|_t^2\right)\right)$$

$$= E_P\left(\exp\left(i\sum_{j=1}^{n}\alpha_j g_j(\xi_{t_j})\right)\right)\exp\left(-\frac{1}{2}\sum_{k,l=1}^{m}\gamma_k\gamma_l R(s_k,s_l)\right)$$

This proves both the assertions. $\square$

Let us recall that $\{X_t, 0 \le t \le T\}$ is a system process with values in a *Polish space* (metric, separable, and complete) $\mathcal{X}$ and assume that the observation process is given by

$$Y_t = \beta(t,X) + N_t, \quad 0 \le t \le T, \tag{5.4}$$

where $\beta(t,X)$ is a non-anticipative function of X_t, $\beta(t,x) \in K(R)$, and $\{N_t, 0 \le t \le T\}$ is independent of $\{X_t, 0 \le t \le T\}$. Let us denote

$$\langle Y,\beta(\cdot,X)\rangle_t = \|\beta(\cdot,X)\|_t^2 + \pi_t(\beta(\cdot,X)). \tag{5.5}$$

With $X = \xi$ and $\beta(\cdot,X) = f(\cdot,\xi)$, we obtain

$$\frac{dP}{dQ_t} = e^{\langle Y,\beta(\cdot,X)\rangle_t - \frac{1}{2}\|\beta(\cdot,X)\|_t^2}, \quad \text{a.e. } Q_t. \tag{5.6}$$

We justify (5.6) by noting that there exists a sequence $\beta^n(\cdot,X) = \sum_{j=1}^{n} a_j(X)R(\cdot,t_j) \in K(R,t)$, $n = 1,2...$, such that $\beta^n(\cdot,X) \to \beta(\cdot,X)$ in $K(R,t)$. Then a.s. Q_t and hence a.s. P,

$$\begin{aligned}\langle Y, \beta(\cdot,X)\rangle_t &= \lim_{n\to\infty} \langle Y, \beta^n(\cdot,X)\rangle_t \\ &= \lim_{n\to\infty} \|\beta^n(\cdot,X)\|_t^2 + \pi_t(\beta^n(\cdot,X)) \\ &= \|\beta(\cdot,X)\|_t^2 + \pi_t(\beta(\cdot,X)),\end{aligned}$$

by continuity of the stochastic integral.

By (5.2) and Theorem 5.1.1, for any F_T^X-measurable integrable function $g(T,X)$ we have

$$\begin{aligned}E_P\left(g(T,X)\big|\mathcal{F}_t^Y\right) &= \frac{E_{Q_t}\left(g(T,X)\frac{dP}{dQ_t}\big|\mathcal{F}_t^Y\right)}{E_{Q_t}\left(\frac{dP}{dQ_t}\big|\mathcal{F}_t^Y\right)} \\ &= \frac{E_{Q_t}\left(g(T,X)\exp\left(\langle Y,\beta(\cdot,X)\rangle_t - \frac{1}{2}\|\beta(\cdot,X)\|_t^2\right)\big|\mathcal{F}_t^Y\right)}{E_{Q_t}\left(\exp\left(\langle Y,\beta(\cdot,X)\rangle_t - \frac{1}{2}\|\beta(\cdot,X)\|_t^2\right)\right)}\end{aligned}$$

Using independence of the processes X and Y under the probability measures Q_t and $Q_t \circ X^{-1} = P \circ X^{-1}$, we have

$$\begin{aligned}E_{Q_t}\left(\Phi(X,Y)\big|\mathcal{F}_t^Y\right)(\omega) &= \int_\Omega \Phi(X(\omega'),Y(\omega))Q_t(d\omega') \\ &= \int_\Omega \Phi(x,Y(\omega))\,dP_X(x).\end{aligned}$$

In conclusion we have the following theorem.

Theorem 5.1.2 (Bayes Formula). *Suppose that the observation process is given by (5.4) and the noise process $\{N_t, 0 \le t \le T\}$ is centered Gaussian with a continuous covariance function R. Let $\beta(\cdot,X(\omega)) \in K(R)$ P-a.s. Then for any $\mathcal{F}_T^X$-measurable and integrable function $g(T,X)$,*

$$E_P\left(g(T,X)\big|\mathcal{F}_t^Y\right) = \frac{\int g(T,\cdot)\exp\left(\langle Y,\beta(\cdot,\cdot)\rangle_t - \frac{1}{2}\|\beta(\cdot,\cdot)\|_t^2\,dP_X\right)}{\int \exp\left(\langle Y,\beta(\cdot,\cdot)\rangle_t - \frac{1}{2}\|\beta(\cdot,\cdot)\|_t^2\,dP_X\right)}, \qquad (5.7)$$

with $\langle Y,\beta(\cdot,\cdot)\rangle_t$ defined as in (5.5).

Example 5.1.1. *Suppose that the noise N_t is of the form*

$$N_t = \int_0^t F(t,u)\,dW_u,$$

where $F(t,\cdot)$ is in $L^2([0,T])$. Then

$$R(t,s) = \int_0^{t\wedge s} F(t,u)F(s,u)\,du.$$

As in Theorem 2.3.1 the RKHS is given by

$$K(R,t) = \left\{ \varphi \,\middle|\, \varphi(s) = \int_0^s F(s,u)\varphi^*(u)\,du, \, , \varphi \in M \right\},$$

where

$$M = \overline{\mathrm{span}}^{L^2([0,T])} \left\{ F(s,u)1_{[0,s]}(u), \, 0 \le s \le t \right\}.$$

The inner product on $K(R,t)$ takes the form

$$\langle \varphi_1, \varphi_2 \rangle_{K(R,t)} = \int_0^t \varphi_1(u)\varphi_2(u)\,du,$$

with $\varphi_i(s) = \int_0^s F(s,u)\varphi^*(u)\,du$, $i = 1,2$.

Suppose now that the observation process is given by

$$Y_t = \int_0^t F(t,u)\tilde{h}(u,X_u)\,du + N_t \tag{5.8}$$

with $\tilde{h}(\cdot,X.) \in \overline{\mathrm{span}}^{L^2([0,T])} \left\{ F(s,\cdot)1_{[0,s]}(\cdot), \, 0 \le s \le t \right\}$. Then for $\varphi \in K(R,t)$,

$$
\begin{aligned}
\langle Y, \varphi \rangle_t &= \int_0^t \varphi^*(u)\tilde{h}(u,X_u)\,du + \int_0^t \varphi^*(u)\,dW_u \\
&= \int_0^t \varphi^*(u)\,d\hat{Y}_u,
\end{aligned}
$$

where $\left\{ \hat{Y}_s = \int_0^s \tilde{h}(u,X_u)\,du + W_s, \, 0 \le s \le T \right\}$.

Hence, we obtain the Bayes formula,

$$E\left(g(T,X)\,\middle|\,\mathcal{F}_t^Y\right) = \frac{\int g(T,x)\exp\left(\int_0^t \tilde{h}(u,X_u)\,d\hat{Y}_u - \frac{1}{2}\int_0^t \left|\tilde{h}(u,X_u)\right|^2 du\right) dP_X}{\int \exp\left(\int_0^t \tilde{h}(u,X_u)\,d\hat{Y}_u - \frac{1}{2}\int_0^t \left|\tilde{h}(u,X_u)\right|^2 du\right) dP_X} \tag{5.9}$$

Exercise 5.1.2. *With the notation of Example 5.1.1, assume that $Y_t = \int_0^t h(u,X_u)\,du + W_t$, where $\{W_t, \, 0 \le t \le T\}$ is a Wiener process, $h(\cdot,X) \in L^2([0,T])$ and $F(t,u) = 1_{[0,t]}(u)$. Show the following Kallianpur–Striebel formula:*

$$E\left(g(T,X)\,\middle|\,\mathcal{F}_t^Y\right) = \frac{\int g(T,x)\exp\left(\int_0^t h(u,X_u)\,dW_u - \frac{1}{2}\int_0^t \left|\tilde{h}(u,X_u)\right|^2 du\right) dP_X}{\int \exp\left(\int_0^t \tilde{h}(u,X_u)\,d\hat{Y}_u - \frac{1}{2}\int_0^t \left|\tilde{h}(u,X_u)\right|^2 du\right) dP_X}$$

Example 5.1.2. *Let $\{X_t, \, 0 \le t \le T\}$ be a continuous process with values in a Polish space $\mathcal{X}$. Suppose the observation process is given by*

$$Y_t = \int_0^t h(X_u)\,du + \tilde{N}_t, \quad 0 \le t \le T,$$

where h is a continuous map from $\mathcal{X}$ to $\mathbb{R}$ and

$$\tilde{N}_t = m_t + \int_0^t \psi(t,s)\,dW_s, \quad 0 \le t \le T. \tag{5.10}$$

Here the function ψ satisfies conditions described below.

Condition 1. *The function $\psi : [0,T] \times [0,T] \to \mathbb{R}$ is continuously differentiable. Denote by $C_{(0)}^r([0,T],\mathbb{R})$ real-valued r-times continuously differentiable functions on $[0,T]$ vanishing at zero. Define an operator $\Psi : C_{(0)}^0([0,T],\mathbb{R}) \to C_{(0)}^0([0,T],\mathbb{R})$ by*

$$(\Psi\varphi)(t) = \Psi(t,t)\varphi(t) - \int_0^t \varphi(s)\frac{\partial\psi(t,s)}{\partial s}\,ds.$$

Note that if $\varphi \in C_{(0)}^1([0,T],\mathbb{R})$, then

$$(\Psi\varphi)(t) = \int_0^t \Psi(t,s)\varphi'(s)\,ds.$$

Let $\mathcal{R}(\Psi) = \left\{ \Psi\varphi \,\middle|\, \varphi \in C_{(0)}^0([0,T],\mathbb{R}) \right\}$. Since for $f,g \in C_{(0)}^0([0,T],\mathbb{R})$ and $0 \le u \le t \le T$

$$(\Psi f)(u) - (\Psi g)(u) = \Psi(u,u)(f(u) - g(u)) - \int_0^u (f(s) - g(s))\frac{\partial\psi(u,s)}{\partial s}\,ds,$$

we conclude that Ψ is a causal operator, that is, $(\Psi f)(u) = (\Psi g)(u)$ for $u \le t$, if $f(s) = g(s)$ for $s \le t$.

Condition 2. *The operator Ψ has causal inverse $S : \mathcal{R}(\Psi) \to C_{(0)}^0([0,T],\mathbb{R})$, such that $S\Psi\varphi = \varphi$ for $\varphi \in C_{(0)}^0([0,T],\mathbb{R})$ and for $g \in C_{(0)}^1([0,T],\mathbb{R}) \cup \mathcal{R}(\Psi)$, Sg is differentiable and $(Sg)' \in L^2[0,T]$.*

Condition 3. *The function m_t is continuously differentiable in t and is an element of $\mathcal{R}(\Phi)$.*

Denote $m_t' = \frac{dm_t}{dt}$ and

$$(Lf)(t) = \frac{d}{dt}(Sg)(t), \tag{5.11}$$

where $g(t) = \int_0^t f(s)\,ds$. Let us observe that

$$Y_t = \int_0^t H(X_s)\,ds + \int_0^t m_s'\,ds + N_t,$$

where $N_t = \int_0^t \Psi(t,s)\,dW_s$. Let

$$R(t,s) = E N_t N_s = \int_0^t \Psi(t,u)\psi(s,u)\,du,$$

then

$$K(R) = \left\{ g \, \middle| \, g(t) = \int_0^t \psi(t,u)g^*(u)\,du, \; g^* \in M \right\},$$

where $M = \overline{\text{span}}^{L^2([0,T])} \left\{ \psi(t,\cdot)1_{[0,t]}(\cdot), \, 0 \le t \le T \right\}$. *We observe that* $M^\perp = \{0\}$
under Conditions 1 and 2. Indeed, let $f \in M^\perp$, *then*

$$\int_0^t \psi(t,s)f(s)\,ds = 0, \quad 0 \le t \le T.$$

With $g(t) = \int_0^t f(s)\,ds$, *we have* $\Psi g = 0$. *This implies that*

$$\int_0^t f(s)\,ds = g(t) = (S\Psi g)(t) = 0, \quad 0 \le t \le T,$$

that is, $f = 0$. *Hence,*

$$K(R) = \left\{ g \, \middle| \, g(t) = \int_0^t \psi(t,u)g^*(u)\,du, \; g^* \in L^2([0,T]) \right\}.$$

Now observe that

$$C_{(0)}^1([0,T],\mathbb{R}) \cap \mathcal{R}(\Psi) \subseteq K(R) \subseteq \mathcal{R}(\Psi). \tag{5.12}$$

For $g \in K(R)$, *and* $f \in C_{(0)}^1([0,T],\mathbb{R}) \cap \mathcal{R}(\Psi)$

$$(Sg)(t) = \int_0^t g^*(u)\,du \quad and \quad f^* = L(f'), \tag{5.13}$$

where the operator L is defined in (5.11). To see that this is true consider $g \in$
$K(R)$, *then*

$$g(t) = \int_0^t \psi(t,s)g^*(s)\,ds.$$

Let $\varphi(t) = \int_0^t g^*(u)\,du$, *then* $\varphi \in C_{(0)}^0([0,T],\mathbb{R})$ *and*

$$
\begin{aligned}
(\Psi\varphi)(t) &= \psi(t,t)\varphi(t) - \int_0^t \frac{\partial \psi(t,s)}{\partial s}\varphi(s)\,ds \\
&= \psi(t,t)\int_0^t g^*(s)\,ds - \int_0^t \left(\frac{\partial \psi(t,s)}{\partial s} \int_0^s g^*(u)\,du \right) ds \\
&= \int_0^t \psi(t,s)g^*(s)\,ds = g(t),
\end{aligned}
$$

using integration by parts. Hence, $g \in \mathcal{R}(\Psi)$ *and for* $g \in K(R)$, $(Sg)(t) =$
$\int_0^t g^*(u)\,du$.

Suppose $f \in C_{(0)}^1([0,T],\mathbb{R}) \cap \mathcal{R}(\Psi)$. *Let* $\varphi \in C_{(0)}^0([0,T],\mathbb{R})$ *be such that*
$\Psi\varphi = f$. *Then by Condition 2, we have*

$$\varphi = S\Psi\varphi = (Sf) \in C_{(0)}^1([0,T],\mathbb{R}) \quad and \quad \varphi' = L(f') \in L^2([0,T]).$$

Consequently,

$$
\begin{aligned}
f(t) &= (\Psi\varphi)(t) = \psi(t,t)\varphi(t) - \int_0^t \varphi(s)\frac{\partial\psi(t,s)}{\partial s}\,ds \\
&= \int_0^t \psi(t,s)\varphi'(s)\,ds,
\end{aligned}
$$

completing the proof of both (5.12) *and* (5.13).

We now obtain Theorem 2.1 in [67].

Corollary 5.1.1. *Let the noise process* $\{\tilde{N}_t,\ 0\le t\le T\}$ *be as in* (5.10) *and let Condition 1 through Condition 3 hold true. Suppose the observation process is given by*

$$
Y_t = \int_0^t h(X_s)\,ds + \tilde{N}_t, \qquad 0\le t\le T.
$$

Let P_X *be the probability distribution of* X *on* $C([0,T])$ *and assume that* $\int_0^{\cdot} h(X_s)\,ds \in \mathcal{R}(\Psi)\ P_X$ *a.e.*

Then for any measurable function g *on* $\mathcal{X}$, *the signal state space, satisfying* $E|g(X_t)| < \infty$,

$$
E\left(g(X_t)\big|\mathcal{F}_t^Y\right) = \frac{\int \alpha_t(x,Y)g(x(t))\,dP_X(x)}{\int \alpha_t(x,Y)\,dP_X(x)},
$$

where

$$
\alpha_t(x,Y) = \exp\left(\int_0^t L(h(x)+m')(s)\,d\hat{Y}(s) - \frac{1}{2}\int_0^t |L(h(x)+m')(s)|^2\,ds\right),
$$

and $\hat{Y}_t = \int_0^t (L(h(x)) + m')(s)\,ds + W_t$.

Proof. Let Ω_0 be a set of full measure P_X and such that $\int_0^{\cdot} h(X_s(\omega))\,ds \in \mathcal{R}(\Psi)$ for all $\omega \in \Omega_0$. Let $\omega \in \Omega_0$ be fixed. Since $h(X_s(\omega))$ is continuous in $s \in [0,T]$,

$$
\int_0^{\cdot} h(X_s(\omega))\,ds \in C^1_{\{0\}} \cap \mathcal{R}(\Psi) \subseteq K(R).
$$

Hence, $\int_0^{\cdot} h(X_s)\,ds \in K(R)\ P_X$ a.s. with $\left(\int_0^{\cdot} h(X_s)\,ds\right)^*(t) = L(h(x))(t)$.

Using similar arguments we conclude that $m \in C^1_{\{0\}} \cap \mathcal{R}(\Psi)$, so that $m \in K(R)$ and $m^* = L(m')$. We obtain

$$
Y_t = \int_0^t L(h(x)+m')(s)\psi(t,s)\,ds + \int_0^t \psi(t,s)\,dW_s.
$$

The corollary follows now from Example 5.1.2 □

Example 5.1.3 (Fractional Brownian Motion)**.** *Suppose the observation process is given by*

$$
Y_t = \int_0^t h(X_u)\,du + W_H(t), \qquad 0\le t\le T,
$$

where $W_H(t)$ is a fBm with covariance

$$R(s,t) = EW_H(t)W_H(s) = \frac{1}{2}\left\{|t|^{2H} + |s|^{2H} - |t-s|^{2H}\right\},$$

where the Hurst parameter $H \in (\frac{1}{2}, 1]$ and W_H is independent of X. We need the following lemma.

Lemma 5.1.1. *Let $\{W_H(t),\ 0 \leq t \leq T\}$ be fBm with Hurst parameter $H \in (\frac{1}{2}, 1]$ and covariance $R(s,t)$. For any continuous function $c : [0, \tau] \to \mathbb{R}\ (\tau > 0)$, suppose g_c^τ satisfies the Carleman equation (see [11])*

$$\int_0^\tau g_c^\tau(u) H(2H-1)|v-u|^{2H-2}\, du = c(v).$$

Let $a : [0, T] \to \mathbb{R}$ be continuous, then $\int_0^\cdot a(u)\, du \in K(R)$ and

$$\pi_t\left(\int_0^\cdot a(u)\, du\right) = \int_0^t g_a^t(u)\, dW_H(u), \qquad (5.14)$$

with

$$\left\|\int_0^\cdot a(u)\, du\right\|_t = \int_0^t g_a^t(u)a(u)\, du. \qquad (5.15)$$

Proof. Consider

$$\int_0^T g_a^T(u)\, dW_H(u) \in H(W_H : T).$$

Then, there exists $\tilde{g} \in K(R)$, such that

$$\begin{aligned}
\tilde{g}(s) &= \langle R(\cdot, s), \tilde{g}(\cdot)\rangle_K (R) \\
&= E\left(W_H \int_0^T g_a^T(s)\, dW_H(s)\right) \\
&= \int_0^s \int_0^T g_a^T(u) H(2H-1)|v-u|^{2H-2}\, du\, dv \\
&= \int_0^s a(v)\, dv, \text{showing (5.14)}
\end{aligned}$$

Hence,

$$\begin{aligned}
E\left(\pi_t\left(\int_0^\cdot a(u)\, du\right)^2\right) &= E\left(\int_0^t g_a^t(u)\, dW_H(u) \int_0^t g_a^t(u)\, dW_H(u)\right) \\
&= \int_0^t \int_0^t g_a^t(u)g_a^t(v) H(2H-1)|u-v|^{2H-2}\, dv\, du \\
&= \int_0^t g_a^t(u)a(u)\, du,
\end{aligned}$$

proving (5.15) $\square$

Consider now $\beta(t,X) = \int_0^t h(X_u)\,du$, $N_t = W_H(t)$, *then we have the following form of the Bayes formula*

$$E\left(f(X_t)\big|\mathcal{F}_t^Y\right) = \frac{\int f(x_t)\exp\left(\int_0^t g_{h(x)}^t(u)\,dY_u - \frac{1}{2}\int_0^t g_{h(x)}^t(u)h(x_u)\,du\right)dP_X(x)}{\int \exp\left(\int_0^t g_{h(x)}^t(u)\,dY_u - \frac{1}{2}\int_0^t g_{h(x)}^t(u)h(x_u)\,du\right)dP_X(x)}.$$

When the signal process is a random variable η independent of the noise process W_H, such that $h(u) = \eta a(u)$, where $a(\cdot)$ is a continuous deterministic function, then using the fact that for a constant k, $g_{ka}^t = kg_a^t$, the Bayes formula can be written as follows:

$$E\left(f(\eta)\big|\mathcal{F}_t^Y\right) = \frac{\int f(x)\exp\left(\int_0^t x g_a^t(u)\,dY_u - \frac{1}{2}x^2\int_0^t g_a^t(u)a(u)\,du\right)dP_\eta(x)}{\int \exp\left(x\int_0^t g_a^t(u)\,dY_u - \frac{1}{2}x^2\int_0^t g_a^t(u)a(u)\,du\right)dP_\eta(x)}.$$

To obtain the result of LeBreton in [71], let us assume that η is Gaussian with mean η_0 and variance γ_0. Then as η is independent of W_H, we know that (η, Y) is jointly Gaussian. Hence, the conditional distribution of η given $\mathcal{F}_t^Y$ is Gaussian with mean $\hat\eta = E\left(\eta\big|\mathcal{F}_t^Y\right)$ and variance $\hat\gamma_t = E\left((\eta - \hat\eta_t)^2\big|\mathcal{F}_t^Y\right)$. Then

$$E\left(e^{\alpha\eta}\big|\mathcal{F}_t^Y\right) = e^{\alpha\hat\eta + \frac{1}{2}\alpha^2\hat\gamma_t}.$$

Taking $f(x) = e^{\alpha x}$ above we also have

$$E\left(e^{\alpha\eta}\big|\mathcal{F}_t^Y\right) \tag{5.16}$$
$$= \frac{\int e^{\alpha x}\exp\left(x\int_0^t g_a^t(u)\,dY_u - \frac{1}{2}x^2\int_0^t g_a^t(u)a(u)\,du\right)\varphi(x;\eta_0,\gamma_0)\,dx}{\exp\left(x\int_0^t g_a^t(u)\,dY_u - \frac{1}{2}x^2\int_0^t g_a^t(u)a(u)\,du\right)\varphi(x;\eta_0,\gamma_0)\,dx},$$

where $\varphi(x;\eta_0,\gamma_0)$ denotes the density of a Gaussian random variable with mean η_0 and variance γ_0. We state the result of LeBreton in [71] as an exercise.

Exercise 5.1.3. (a) *Prove that the numerator in (5.16) is equal to*

$$\sqrt{\gamma_0^{-1}\gamma_t}\exp\left(-\frac{1}{2}\gamma_0^{-1}\eta_0^2 + \frac{1}{2}\gamma_t(\alpha + m_t)^2\right),$$

with $\gamma_t^{-1} = \gamma_0^{-1} + \int_0^t g_a^t(u)a(u)\,du$ and $m_t = \gamma_0^{-1}\eta_0 + \int_0^t g_a^t(u)\,dY_u$.
(b) *Put $\alpha = 0$ in (a) to obtain the denominator.*
(c) *Show that*

$$E\left(e^{\alpha\eta}\big|\mathcal{F}_t^Y\right) = e^{\frac{1}{2}\gamma_t\alpha(\alpha + 2m_t)},$$

and hence

$$\hat\eta_t = \gamma_t\left(\gamma_0^{-1}\eta_0 + \int_0^t g_a^t(u)\,dY_u\right)$$
$$\hat\gamma_t = \gamma_t.$$

The next exercise contains results from Section 3 in [34].

Exercise 5.1.4. *Let us study the observation process*

$$Y_t = \int_0^t h_\alpha(u,X)\,du + N_t,$$

where

$$N_t = \int_0^t \frac{\sigma}{\beta}\left(1 - e^{-\beta(t-s)}\right) dW_s$$

is the Ornstein–Uhlenbeck process and $h_\alpha(\cdot,x) \in L^2([0,T],dt)$. Assume that the derivative $h'_\alpha(t,x)$ with respect to t exists and $\int_0^t |h_\alpha(s,x) - h(x_s)|^2\,ds \to 0$ as $\alpha \to \infty$.

(a) *Check that $h_\alpha(t,x) = \alpha \int_{t-1/\alpha}^t h(x_s)\,ds$ satisfies the above conditions for $h(x.) \in L^2([0,T],dt)$.*

(b) *Show that if*

$$\tilde{h}_{\alpha,\beta,\sigma}(t,x) = \frac{\sigma}{\beta}h_\alpha(t,x) + \frac{1}{\sigma}h'_\alpha(t,x),$$

then

$$\int_0^t h_\alpha(s,x)\,ds = \frac{\sigma}{\beta}\int_0^t \left(1 - e^{-\beta(t-u)}\right)\tilde{h}_{\alpha,\beta,\sigma}(u,x)\,du,$$

giving

$$Y_t^{\alpha,\beta,\sigma} = \frac{\sigma}{\beta}\int_0^t \left(1 - e^{-\beta(t-u)}\right)\tilde{h}_{\alpha,\beta,\sigma}(u,X)\,du + N_t.$$

(c) *Using part (b) and (5.9) prove that for a bounded real-valued measurable function f*

$$E\left(f(X_t)\,\Big|\,\mathcal{F}_t^{Y^{\alpha,\beta,\sigma}}\right) = \frac{\int f(x_t)q_t^{\alpha,\beta,\sigma}\left(x,Y^{\alpha,\beta,\sigma}\right)dP_X(x)}{\int q_t^{\alpha,\beta,\sigma}\left(x,Y^{\alpha,\beta,\sigma}\right)dP_X(x)},$$

where

$$q_t^{\alpha,\beta,\sigma}\left(x,Y^{\alpha,\beta,\sigma}\right) = \exp\left(\int_0^t \left(\frac{\beta}{\sigma}h_\alpha(u,x) + \frac{1}{\sigma}h'_\alpha(u,x)\right)d\hat{Y}_u^{\alpha,\beta,\sigma}\right.$$
$$\left. - \frac{1}{2}\int_0^t \left(\frac{\beta}{\sigma}h_\alpha(u,x) + \frac{1}{\sigma}h'_\alpha(u,x)\right)^2 du\right)$$

$$\hat{Y}_t^{\alpha,\beta,\sigma} = \int_0^t \left(\frac{\beta}{\sigma}h_\alpha(u,X) + \frac{1}{\sigma}h'_\alpha(u,X)\right)du + W_t.$$

(c) *Let*

$$dP'^{\alpha,\beta,\sigma} = 1/q_t\left(X,Y^{\alpha,\beta,\sigma}\right)dP.$$

Then under the measure $P'^{\alpha,\beta,\sigma}$ the process $Y^{\alpha,\beta,\sigma}$ is a Brownian motion,

the distribution of X remains the same as under the measure P, and X is
independent of $Y^{\alpha,\beta,\sigma}$. *Denote by* W^Q *the coordinate process. Consider*

$$p_t^{\alpha,\beta,\sigma}(x,W^Q) = \exp\left\{ \int_0^t \left(\tfrac{\beta}{\sigma}h_\alpha(u,x) + \tfrac{1}{\sigma}h'_\alpha(u,x) \right) dW_u^Q \right.$$
$$\left. -\tfrac{1}{2} \int_0^t \left(\tfrac{\beta}{\sigma}h_\alpha(u,x) + \tfrac{1}{\sigma}h'_\alpha(u,x) \right)^2 du \right\},$$

and, using pathwise formula for stochastic integrals in [56], show that

$$p_t^{\alpha,\beta,\sigma}\left(x,\hat{Y}^{\alpha,\beta,\sigma}\right) = q_t^{\alpha,\beta,\sigma}\left(x,Y^{\alpha,\beta,\sigma}\right) \quad P\text{–}a.s.$$

(e) *Use the fact that*

$$\int_0^T |h_\alpha(u,x) - h(x_u)|^2 du \to 0, \text{ as } \alpha \to \infty,$$

to show that as β, $\sigma \to \infty$ *with* $\beta/\sigma \to 1$, *and then* $\alpha \to \infty$,

$$p_t^{\alpha,\beta,\sigma}\left(x,W^Q\right) \to p_t(x,W^Q) = \exp\left\{ \int_0^t h(x_u)\,dW_u^Q - \frac{1}{2}\int_0^t h^2(x_u)\,du \right\},$$

a.e. with respect to the measure Q, over a subsequence, where Q is the stan-
dard Wiener measure on $C([0,T],\mathbb{R})$. *Conclude from the Bayes formula in*
part (c), using Scheffe's theorem, that

$$E\left(f(X_t)\Big|\mathcal{F}_t^{Y^{\alpha,\beta\sigma}}\right) \to E\left(f(X_t)|\mathcal{F}_t^{Y_t}\right),$$

where $Y_t = \int_0^t h(X_s)\,ds + W_t$.

5.2   Zakai Equation

Recall that the main goal of filtering theory is to find the conditional expectation of a function of the system process $f(X_t)$, given the information contained
in the observation process, $\mathcal{F}_t^Y$, for a large enough class of functions f. In case
of $N_t = W_t$, the Brownian motion, a formula for this conditional expectation was
given by Kallianpur and Striebel in [52]. It was shown in [53], that from this
formula one can obtain the, so called, *FKK equation* (see [52]) for the filter.
However, a simpler equation, called the Zakai equation, [72]), can be obtained
for the un-normalized measure in the numerator of the Bayes formula, and is
basic in obtaining the expectation recursively.

In this section we study an extension of the Zakai equation in case the
observation process is given by

$$Y_t = \int_0^t F(t,u)\tilde{h}(u,X_u)\,du + N_t, \text{ with } N_t = \int_0^t F(t,u)\,dW_u.$$

Denote by $C_c^2(\mathbf{R}^n)$ the space of continuous functions on $\mathbf{R}^n$ with compact support and having bounded derivatives of the order up to two. Define an operator L by

$$(L_t f)(x) = \sum_{j=1}^{n} b_j(t,x)\frac{\partial f}{\partial x_j}(x) + 1/2 \sum_{i,j=1}^{n} \sigma_{i,j}(t,x)\frac{\partial^2 f}{\partial x_i x_j}(x), \qquad (5.17)$$

for $f \in C_c^2(\mathbf{R}^n)$, with $b_j(t,x)$ and $\sigma_{i,j}(t,x)$ bounded and continuous. We assume that $\{X_t, t \in [0,T]\}$ is a *solution* of a *martingale problem*, that is, for each $f \in C_c^2(\mathbf{R}^n)$

$$f(X_t) - \int_0^t (L_u f)(X_u)\, du$$

is an $\mathcal{F}_t^X$ martingale with respect to the measure P. Consider a product probability space $(\Omega \times \Omega, \mathcal{F} \otimes \mathcal{F}, P \otimes P')$, where P' is a probability measure in the Bayes formula (5.9), under which $\hat{Y}$ is a Brownian motion, that is,

$$dP' = \exp\left(-\int_0^t \tilde{h}(u,X_u)\,d\hat{Y}_u + 1/2\int_0^t \left|\tilde{h}(u,X_u)\right|^2 du\right)dP.$$

Then under the measure P', the process $\hat{Y}$ is a Brownian motion independent of X, and $P_X = P_X'$.

Let us now introduce some notation. Given $(\omega', \omega) \in \Omega \times \Omega$ and a function ϕ on Ω, we denote $\phi'(\omega', \omega) = \phi(\omega')$ and $\phi(\omega', \omega) = \phi(\omega)$. For example, $X_t'(\omega', \omega) = X_t(\omega')$ and $X_t(\omega', \omega) = X_t(\omega)$ are independent copies of X_t under the measure $\pi = P \otimes P'$.

Define

$$\alpha_t(\omega', \omega) = \exp\left(\int_0^t \tilde{h}(s, X_s(\omega'))\,d\hat{Y}_s(\omega) - 1/2\int_0^t \left|\tilde{h}(s, X_s(\omega'))\right|^2 ds\right).$$

With this notation, by the Bayes formula (5.9), for any g, $\mathcal{F}_T^X$-measurable on $(\Omega, \mathcal{F}, P)$, we have

$$E\left(g\middle|\mathcal{F}_t^Y\right) = \frac{\int g(\omega')\alpha_t(\omega', \omega)\,dP_X(\omega')}{\int \alpha_t(\omega', \omega)\,dP_X(\omega')}.$$

Define for all $f \in C_c^2(\mathbf{R}^n)$

$$\hat{\sigma}_t\left(f, \hat{Y}\right)(\omega) = \int f(X_t(\omega'))\alpha_t(\omega', \omega)\,dP(\omega').$$

We will need the following lemma (see Theorem 5.14 in [74]),

Lemma 5.2.1. *Let $\{X_t, 0 \le t \le T\}$ be an $\mathcal{F}_t$-martingale,*

$$X_t = \int_0^t a_s\,dW_s, \qquad P\left(\int_0^T a_s^2\,ds < \infty\right) = 1.$$

If $E|a_s| < \infty$, $0 \le s \le T$ and

$$P\left(\int_0^T E\left(|a_s| \big| \mathcal{F}_s^W \right)^2 ds < \infty \right) = 1$$

then P–a.e. for all t,

$$E\left(\int_0^t a_s \, dW_s \big| \mathcal{F}_t^W \right) = \int_0^t E\left(a_s \big| \mathcal{F}_s^W \right) dW_s.$$

We can now state a theorem on the *Zakai equation*.

Theorem 5.2.1 (Zakai Equation). *The quantity $\hat{\sigma}_t(f, \hat{Y})$ defined above satisfies the Zakai equation*

$$d\hat{\sigma}_t\left(f(\cdot), \hat{Y} \right) = \hat{\sigma}_t\left(L_t f(\cdot), \hat{Y} \right) dt + \hat{\sigma}_t\left(\tilde{h}(t, \cdot) f(\cdot), \hat{Y} \right) d\hat{Y}_t. \qquad (5.18)$$

Proof. Let $g_t = f(X_T) - \int_t^T (L_s f)(X_s) ds$. Since f and $L_s f$ are uniformly bounded, $|g_t| \le C$ for some constant C.

Using the fact that $f(X_t) - \int_t^T (L_s f)(X_s) ds$ is a martingale, we obtain that

$$E_P\left(g_t \big| \mathcal{F}_t^X \right) = f(X_t), \quad 0 \le t \le T. \qquad (5.19)$$

For each ω, $\alpha_t(\omega', \omega)$, as a function of ω', is $\mathcal{F}_t^{X'}$-measurable. Now equation (5.19) implies that

$$\begin{aligned}
\hat{\sigma}_t\left(f, \hat{Y} \right) &= \int f\left(X_t(\omega') \right) \alpha_t(\omega', \omega) \, dP(\omega') \\
&= \int E_P\left(g_t(\omega') \alpha_t(\omega', \omega) \big| \mathcal{F}_t^{X'} \right) dP(\omega') \\
&= E_P\left(g_t(\omega') \alpha_t(\omega', \omega) \right) \\
&=: \sigma_t'\left(g_t, \hat{Y} \right).
\end{aligned}$$

By the definition of g_t,

$$dg_t = (L_t f)\left(X_t' \right) dt,$$

and, using the Itô formula,

$$d\alpha_t = \alpha_t \tilde{h}\left(t, X_t' \right) d\hat{Y}_t,$$

as under the measure $\pi = P \otimes P'$, $\hat{Y}_t$ is a Brownian motion.

Hence, using the Fubini theorem, we calculate $\sigma_t'\left(g_t, \hat{Y} \right) = E_P(g_t' \alpha_t)$ as follows:

$$E_P g_0' \alpha_0 + E_P \int_0^t (L_s f)\left(X_s' \right) \alpha_s \, ds + E_P \int_0^t g_s' \alpha_s \tilde{h}\left(s, X_s' \right) d\hat{Y}_s(\omega)$$

$$= E_P g_0' + \int_0^t E_P\left[(L_s f)\left(X_s' \right) \right] \alpha_s \, ds + A_t$$

$$= E_P g_0' + \int_0^t \hat{\sigma}_s \left(L_s f, \hat{Y} \right) ds + A_t,$$

where we denoted

$$A_t = \int \int_0^t g_s(\omega') \alpha_s(\omega', \omega) \tilde{h}\left(s, X_s(\omega')\right) d\hat{Y}_s(\omega) dP(\omega').$$

By the definition of π,

$$\sigma_t'\left(g_t, \hat{Y}\right) = E_\pi \left(g_t' \alpha_t \middle| \mathcal{F}_t^{\hat{Y}} \right).$$

Using Lemma 5.2.1 and the fact that under the measure π, the process $\hat{Y}_t$ is a Brownian motion, we obtain, with $E = E_\pi$,

$$
\begin{aligned}
A_t &= E \left(E \left(\int_0^t g_s' \alpha_s(\omega', \omega) \tilde{h}\left(s, X_s(\omega')\right) d\hat{Y}_s \middle| \mathcal{F}_s^{X', \hat{Y}} \right) \middle| \mathcal{F}_s^{\hat{Y}} \right) \\
&= \int_0^t E \left(E \left(g_s' \alpha_s(\omega', \omega) \tilde{h}\left(s, X_s(\omega')\right) \middle| \mathcal{F}_s^{X', \hat{Y}} \right) \middle| \mathcal{F}_s^{\hat{Y}} \right) d\hat{Y}_s \\
&= \int_0^t E \left(\alpha_s(\omega', \omega) \tilde{h}\left(s, X_s(\omega')\right) E \left(g_s(\omega') \middle| \mathcal{F}_s^{X', \hat{Y}} \right) \middle| \mathcal{F}_s^{\hat{Y}} \right) d\hat{Y}_s \\
&= \int_0^t \hat{\sigma}_s \left(\tilde{h}\left(s, X_s(\omega')\right) f \left(X_s(\omega')\right), \hat{Y} \right) d\hat{Y}_s,
\end{aligned}
$$

completing the proof.   □

Remark 5.2.1. *Equation (5.18) holds for all functions f in the domain of L_t.*

Exercise 5.2.1. **(a)** *Derive the Zakai equation for the case $Y_t = \int_0^t h(X_u) du + W_t$.*

(b) *Let $N_t = W_H(t)$ and $\beta(t, X) = \int_0^t h(X_u) du$.*
Use the form of $E\left(f(X_t) \middle| \mathcal{F}_t^Y \right)$ given in Section 5.1 and an argument as in the proof of Theorem 5.2.1 to derive the Zakai equation for $\int f(X_t(\omega')) \alpha_t(\omega, \omega') dP \circ X^{-1}(\omega)$.

(c) *Let $N_t = M_t$ be a centered Gaussian martingale and $m(t) = EM_t^2$. Show that*
  (i) *$K(R) = \{g | g(t) = \int_0^t g^*(u) dm(u), 0 \le s \le T, g^* \in L^2(dm)\}$ and consequently, $K(R, t) = \{g | g(s) = \int_0^s g^*(u) dm(u), 0 \le s \le T, g^* \in L^2(dm)\}$.*
  (ii) *With the observation process $Y_t = \int_0^t f(s, X) dm(s) + M_t$ compute the Bayes formula for $E\left(g(T, X) \middle| \mathcal{F}_t^Y \right)$.*
  (iii) *Let X be a solution to a martingale problem with a generator L_t. Find the Zakai equation for $\hat{\sigma}_t(f, Y)$.*
  (iv) *Use the Itô formula to find the FKK equation for*

$$\hat{\Pi}_t(f) = \frac{\hat{\sigma}(f, Y)}{\hat{\sigma}(v, Y)},$$

where $v = Y_t - \int_0^t \hat{\Pi}_s(h) dm(s)$. Show that

$$d\hat{\Pi}_t(f) = \hat{\Pi}_t (L_t f) \, dt + \left(\hat{\Pi}_t(hf) - \hat{\Pi}_t(f) \hat{\Pi}_t(h) \right) dv_t. \tag{5.20}$$

5.3   Kalman Filtering for Fractional Brownian Motion Noise

Let us define for $0 < s < t \leq T$,

$$k_H(t,s) = \kappa_H^{-1} s^{1/2-H} (t-s)^{1/2-H}, \text{ where } \kappa_H = 2H\Gamma(3/2-H)\Gamma(H+1/2),$$
$$w_t^H = \lambda_H^{-1} t^{2-2H}, \text{ with } \lambda_H = \frac{2H\Gamma(3-2H)\Gamma(H+1/2)}{\Gamma(3/2-H)},$$

and a process

$$M_t^H = \int_0^t k_H(t,s)dW_s^H.$$

The integral with respect to fBm W_t^H is described in [93]. The process $\{M_t^H, 0 \leq t \leq T\}$ is a Gaussian martingale. Define

$$Q_H^C(t) = \frac{d}{dw_t^H} \int_0^t k_H(t,s)C(s)\,ds, \tag{5.21}$$

where $C(t)$ is an $\mathcal{F}_t$-adapted process and the derivative is understood to be the Radon-Nikodym derivative.

Let $Y_t = \int_0^t C(s,X)\,ds + W_t^H$. Then one can show, as in [58], that

$$Z_t = \int_0^t Q_H^C(s)\,dw_s^H + M_t^H$$

is an $\mathcal{F}_t^Y$ semi-martingale and $\mathcal{F}_t^Y = \mathcal{F}_t^Z$. Then the filtering problem for the observation process Y_t is equivalent to the filtering problem for the system process X_t and the observation process

$$Z_t = \int_0^t Q_H^C(s,X)\,dw_s^H + M_t^H. \tag{5.22}$$

By Exercise 5.2.1 the FKK equation (5.20) for $\hat{\Pi}$ reduces to

$$d\hat{\Pi}_t(f) = \hat{\Pi}_t(L_t f)dt + \left[\hat{\Pi}_t(Q_H^C f) - \hat{\Pi}_t(f)\hat{\Pi}_t(Q_H^C)\right]d\nu_t, \tag{5.23}$$

where

$$\nu_t = Z(t) - \int_0^t \hat{\Pi}_s(Q_H^C)\,dw_s^H \tag{5.24}$$

is a continuous Gaussian $\mathcal{F}_t^Y$-martingale with variance w_t^H by Theorem 2 in [58], which we state as the next lemma.

Lemma 5.3.1. *Suppose that the sample paths of Q_H^C defined in (5.21) belong P-a.s. to $L^2([0,T], dw^H)$. Let the processes $Z(t)$ and ν_t be defined by (5.22) and (5.24). Then*

(a) *the process ν_t is a continuous Gaussian $\mathcal{F}_t^Y$-martingale with the variance function w_t^H.*

(b) *If N_t is a square integrable $\mathcal{F}_t^Y$-martingale, $N_0 = 0$, then there exists a $\mathcal{F}_t^Y$-adapted process Φ_t, such that $E \int_0^T \Phi_t^2 \, dw_t^H < \infty$ and P-a.s.*

$$N_t = \int_0^t \Phi_s \, dv_s, \quad t \in [0,T].$$

Consider now the *Kalman filtering problem*. Let the system process and observation processes be given by

$$X_t = \int_0^t b(u) X_u \, du + \int_0^t \sigma(u) \, dW_u$$

$$Y_t = \int_0^t c(u) X_u \, du + W_t^H,$$

where the processes W_t and W_t^H are independent. Because (X_t, Z_t) is jointly Gaussian, we obtain with $s \leq t$,

$$\hat{\Pi}_t(X_t X_s) - \hat{\Pi}_t(X_t)\hat{\Pi}_t(X_s)$$
$$= E\left\{ (X_t - \hat{\Pi}_t(X_t))(X_s - \hat{\Pi}_t(X_s)) \big| \mathcal{F}_t^Y \right\}$$
$$= E\left\{ (X_t - \hat{\Pi}_t(X_t))(X_s - \hat{\Pi}_t(X_s)) \right\},$$

and a similar expression for $t \leq s$. Let us denote

$$\Gamma(t,s) = E\left\{ (X_t - \hat{\Pi}_t(X_t))(X_s - \hat{\Pi}_t(X_s)) \right\}.$$

Then by (5.23), with $f(x) = x$ and $L_t f = b(t)xf'(x) + \sigma^2(t)f''(x)$, we obtain the following equation:

$$d\hat{\Pi}_t(X_t) = b(t)\hat{\Pi}_t(X_t) \, dt + \left(\int_0^t k_H(t,s)\Gamma(t,s) \, ds \right) dv_t. \qquad (5.25)$$

Denote by $\gamma(t) = EX_t^2$, and $F(t) = E\left(\hat{\Pi}_t^2(X_t) \right)$. Then using the Itô formula for $f(x) = x^2$ and taking the expectation, we arrive at

$$d\gamma(t) = 2b(t)\gamma(t) dt + \sigma^2(t) dt$$

and

$$dF(t) = 2b(t)F(t) dt + \left(\int_0^t k_H(t,s)\Gamma(t,s) \, ds \right)^2 dW_t^H.$$

Let us consider

$$\Gamma(t,t) = E(X_t - \hat{\Pi}(X_t))^2$$
$$= E(X_t^2) - E(\hat{\Pi}_t^2(X_t))$$
$$= \gamma(t) - F(t).$$

Then we obtain the following equation for $\Gamma(t,t)$:

$$d\Gamma(t,t) = 2b(t)\Gamma(t,t)dt + \sigma^2(t)dt - \left(\int_0^t k_H(t,s)\Gamma(t,s) \, ds \right)^2 dw_s^H. \quad (5.26)$$

Note that for $H = \frac{1}{2}$ Equations (5.25) and (5.26) reduce to Kalman equations in the case of Brownian motion.

Chapter 6

Equivalence and Singularity

In this chapter, we first present general conditions for two probability measures to be absolutely continuous or singular on a σ-field $\mathcal{F}$ of a set Ω, where the two measures are defined. The main problem we want to study is obtaining conditions on covariances of two Gaussian processes $\{X_t, t \in T\}$, and $\{Y_t, t \in T\}$ in order for them to be singular as inference involves the existence of a likelihood ratio. As an application of current interest, we obtain some results in [120] on the interpolation error and sufficient conditions on partially observed stationary random fields.

6.1 General Problem

We begin with general conditions for equivalence and singularity. Let $\mathcal{F}$ be a σ-field of subsets of a set Ω and P_0, P_1 be two probability measures defined on $\mathcal{F}$, then the Lebesgue decomposition of P_1 with respect to P_0 can be written in the following form ([110], p. 211):

$$P_1(A) = \int_A \frac{p_1}{p_0} \, dP_0 + P_1 (A \cap \{p_0 = 0\}), \quad \text{for all } A \in \mathcal{F},$$

where p_0, p_1 are Radon-Nikodym densities of, respectively, P_0 and P_1, with respect to $P_0 + P_1$.

As an immediate consequence we obtain conditions for singularity and equivalence of P_0 and P_1,

(i) $P_0 \perp P_1$ on $\mathcal{F}$ if and only if $P_0 \left(\frac{p_1}{p_0} = 0 \right) = 1$,

(ii) $P_0 \ll P_1$ on $\mathcal{F}$ if and only if $P_0 \left(\frac{p_1}{p_0} > 0 \right) = 1$.

Exercise 6.1.1. *Prove the equivalence (i) above.*

To prove the equivalence (ii), we observe that under the condition $P_0 \left(\frac{p_1}{p_0} > 0 \right) = 1$, for any $A \in \mathcal{F}$, $P_0(A) > 0$ implies $P_1(A) > 0$, giving that $P_0 \ll P_1$. On the other hand, if $P_0 \ll P_1$, then $P_1(p_0 = 1) = 0$ implies $P_0(p_0 = 1) = 0$, therefore $P_0(0 < p_0 < 1) = 1$. In addition, $p_0 + p_1 = 1$ a.e. with respect to $P_0 + P_1$, giving that $p_1 > 0$ P_0-a.e. Hence, $P_0 \left(\frac{p_1}{p_0} > 0 \right) = 1$.

Let us denote by $\rho = \frac{p_1}{p_0}$ the Radon-Nikodym density of the absolutely continuous part of P_1 with respect to P_0. We note that $\rho \geq 0$ and $\lim_{\alpha \to 0} \rho^{\alpha} = 1_{\rho > 0}$

P_0–a.e. Since for $\alpha_0 < 1$

$$E_{P_0}(\rho^\alpha)^{\frac{1}{\alpha_0}} < \infty$$

for all $0 < \alpha \leq \alpha_0$, we conclude that the family of functions $\{\rho^\alpha, 0 < \alpha < \alpha_0\}$ is uniformly integrable. Hence,

$$\lim_{\alpha \to 0} E_{P_0}(\rho^\alpha) = P_0(\rho > 0).$$

Thus, we obtain the following conditions for equivalence and singularity:

(i) $P_0 \perp P_1$ on $\mathcal{F}$ if and only if $E_{P_0}\rho^\alpha = 0$ for all $0 < \alpha < 1$

(ii) $P_0 \ll P_1$ on $\mathcal{F}$ if and only if $\lim_{\alpha \to 0} E_{P_0}\rho^\alpha = 1$ (6.1)

To study the likelihood ratio of a process we note that $\mathcal{F}_T = \sigma\left(\bigcup_{i\in I}\mathcal{F}_i\right)$, where I denotes the family of finite subsets of T. Let us consider now I as a partially ordered set. Then $\{\mathcal{F}_i, i \in I\}$ is a non-decreasing family of sub σ-fields of $\mathcal{F}$, such that $\mathcal{F} = \sigma\left(\bigcup_{i\in I}\mathcal{F}_i\right)$. Let ρ_i denote the density of the absolutely continuous part of $P1$ with respect to P_0 on $\mathcal{F}_i$. It is left for the reader as an exercise to examine the proof of the Radon–Nikodym theorem and conclude that

$$\int_A \rho_i \, dP_0 \geq \int_A \rho_j \, dP_0 \quad \text{for all } A \in \mathcal{F}_i, \, j \geq i. \tag{6.2}$$

Exercise 6.1.2. *Prove* (6.2).

But the inequality (6.2) implies that $\{\rho_i, i \in I\}$ is a supermartingale. By Jensen's inequality, for $0 < \alpha < 1$,

$$E_{P_0}^{\mathcal{F}_i}\left(\rho_j^\alpha\right) \leq \rho_i^\alpha, \quad j \geq i,$$

giving that $E_{P_0}\rho_i^\alpha$ is non-increasing in i for each α. We now obtain conditions, in terms of ρ_i, for the measures P_0 and P_1 to be absolutely continuous or singular.

Lemma 6.1.1. *Let Ω be a set and $\mathcal{F}$ be a σ-field of its subsets, such that $\mathcal{F} = \sigma\left(\bigcup_{i\in I}\mathcal{F}_i\right)$. Let P_0, P_1 be probability measures on $(\Omega, \mathcal{F})$ and ρ_i, ρ be the Radon–Nikodym densities of the absolutely continuous part of P_1 with respect to P_0 on $\mathcal{F}_i$ and $\mathcal{F}$, respectively. Then*

$$\inf_{i \in I} E\rho_i^\alpha = E\rho^\alpha, \quad \text{for all } 0 < \alpha < 1.$$

Proof. Since ρ is $\mathcal{F}$ measurable, there exists a sequence $i_1 < i_2 < ...$ of elements of I, such that ρ is measurable with respect to $\sigma\left(\bigcup_{k=1}^{\infty}\mathcal{F}_{i_k}\right)$. It is known (see [92], pg. 41), that $\rho_{i_k} \to \rho$, P_0–a.e. By the uniform integrability of the family $\left\{\rho_{i_k}^\alpha, k = 1, 2...\right\}$ for $0 < \alpha < 1$, we have that

$$E\rho^\alpha = \lim_{k \to \infty} E\rho_{i_k}^\alpha \geq \inf_{i \in I} \geq E\rho^\alpha.$$

Observe that $\inf_{i \in I} E\rho_i^\alpha = \lim_{i \in I} E\rho_i^\alpha$. $\quad\square$

Theorem 6.1.1. *Let P_0 and P_1 be two probability measures on a measurable space $(\Omega, \mathcal{F})$ and I be a directed family of sets. If $\{\mathcal{F}_i, i \in I\}$ is a nondecreasing family of sub σ-fields of $\mathcal{F}$, such that $\mathcal{F} = \sigma(\bigcup_{i \in I} \mathcal{F}_i)$, and ρ_i are the Radon–Nikodym densities of the absolutely continuous part of P_1 with respect to P_0 on $\mathcal{F}_i$, then*

(i) *$P_0 \perp P_1$ on $\mathcal{F}$ if and only if $\lim_{i \in I} \int \rho_i^\alpha \, dP_0 = 0$, $0 < \alpha < 1$.*

(ii) *$P_0 \ll P_1$ on $\mathcal{F}$ if and only if for every $\varepsilon > 0$, there exists $\alpha(\varepsilon) \in (0,1)$ such that $\int \rho_i^\alpha > 1 - \varepsilon$ for all $\alpha \in (0, \alpha(\varepsilon)]$ and all $i \in I$.*

Exercise 6.1.3. (a) *Let $\mathcal{F}_1 \subseteq \mathcal{F}_2 \subseteq \dots$ be sub σ-fields of subsets of Ω and $\mathcal{F} = \sigma(\bigcup_{n=1}^\infty \mathcal{F}_n)$. Let P and Q be two measures on $\mathcal{F}$, such that $Q \equiv P$ on $\mathcal{F}_n$ for each n, and let $\rho_n = \frac{dQ|\mathcal{F}_n}{dP|\mathcal{F}_n}$. Let $\mathcal{G} = \bigcap_{n=1}^\infty \sigma\left\{\frac{\rho_{k+1}}{\rho_k}, k \geq n\right\}$. If $\mathcal{G}$ is trivial (i.e., for all $A \in \mathcal{G}$, $P(A) = 0$ or 1), then $P \equiv Q$ or $P \perp Q$, and $P \equiv Q$ on $\mathcal{F}$ if and only if $E_P \rho_n^\alpha \not\to 0$ for some α.*

(b) *With the same notation as above, show that $P \perp Q$ if and only if $\lim_{n \to \infty} E \rho_n^{1/2} = 0$.*

(c) *(Kakutani Theorem) Let $\Omega = \prod_{k=1}^\infty \Omega_k$, $\mathcal{F} = \otimes_{k=1}^\infty A_k$, $P_0 = \otimes_{k=1}^\infty \mu_k$, and $P_1 = \otimes v_k$, where μ_k, v_k are probability measures on (Ω_k, A_k). Let $\tau_i : \Omega \to \prod_{k=1}^i \Omega_k$ and $\mathcal{F}_i = \tau_i^{-1}\left(\otimes_{i=1}^i A_k\right)$. Clearly $\mathcal{F} = \sigma(\bigcup_{i=1}^\infty \mathcal{F}_i)$ and $\mathcal{F}_i$ are non-decreasing σ-fields. Assume that $v_k \equiv \mu_k$ for each k, and $\rho_i(\omega) = \prod_{k=1}^i \frac{d\mu_k}{dv_k}(\omega_k)$, where $\omega = (\omega_1, \omega_2, \dots) \in \Omega$.*

(i) *Show that the σ-field $\mathcal{G}$ defined as in part (a) is trivial.*

(ii) *$P \equiv Q$ if and only if the following product converges to a positive limit, $\prod_{k=1}^\infty \int_\Omega \frac{d\mu_k}{dv_k} > 0$.*

(d) *Assume that in part (c) the measures $v_k = G(0,1)$ are Gaussian with mean zero and variance one, and $\mu_k = G(a_k, 1)$. Then*

$$\frac{d\mu_k}{dv_k}(u) = e^{a_k u - \frac{1}{2} a_k^2}.$$

Compute $E\left(\frac{d\mu_k}{dv_k}\right)^{1/2} dv_k$ and show that $P_0 \equiv P_1$ if and only if $\{a_k\}_{k=1}^\infty \in l_2$.

The natural question is if we define measures $P_0 = \otimes_{k=1}^\infty \mu$ and $P_{\underline{a}} = \otimes_{k=1}^\infty \mu_{a_k}$, where $\underline{a} = (a_1, a_2, \dots) \in \mathbb{R}^\infty$ and $\mu_{a_k}(B) = \mu(B + a_k)$ for $B \in \mathcal{B}(\mathbb{R})$, are there non-Gaussian measures μ for which the set

$$E_P = \left\{\underline{a} \in \mathbb{R}^\infty : P_{\underline{a}} \equiv P\right\} = l_2?$$

We shall study this problem now. Assume $\mu(dx) = p(x)dx$ and $p(x) > 0$. Then

$$\frac{d\mu_k}{d\mu}(x) = \frac{p(x + a_k)}{p(x)}.$$

Hence, by Kakutani's theorem $P_{\underline{a}} \equiv P$ if and only if

$$\prod_{k=1}^{\infty} \int_{-\infty}^{\infty} p^{\frac{1}{2}}(x)p^{\frac{1}{2}}(x+a_k)\,dx$$

converges to a positive limit. Since $p^{\frac{1}{2}}(x) \in L^2(\mathbb{R},dx)$, we can define the Fourier–Plancherel transform of $h(x) = p^{\frac{1}{2}}(x)$ by

$$\hat{h}(u) = (2\pi)^{-\frac{1}{2}} \int_{-\infty}^{\infty} e^{iux}h(x)\,dx.$$

Then using the Plancherel theorem

$$\prod_{k=1}^{\infty} \int_{-\infty}^{\infty} p^{\frac{1}{2}}(x)p^{\frac{1}{2}}(x+a_k)\,dx = \prod_{k=1}^{\infty} \int_{-\infty}^{\infty} e^{-iua_k}\left|\hat{h}(x)\right|^2\,du. \qquad (6.3)$$

Let $\beta(u) = \left|\hat{h}(x)\right|^2$, then $\int_{-\infty}^{\infty} \beta(u)\,du = 1$. Taking real parts in Equation (6.3) we conclude that $\underline{a} \in E_P$ if and only if $\prod_{k=1}^{\infty} \int_{-\infty}^{\infty} \cos(a_k u)\beta(u)\,du > 0$. Hence,

$$E_P = \left\{\underline{a} : \sum_{k=1}^{\infty} \int_{-\infty}^{\infty} (1-\cos(a_k u))\beta(u)\,du < \infty\right\}.$$

Using the inequality $|1-\cos(a_k u)| \le \frac{a_k^2 u^2 \wedge 1}{2}$ we define a function

$$\psi(\lambda) = \int_0^{\infty} \left(\lambda^2 u^2 \wedge 1\right)\beta(u)\,du.$$

Then the function ψ satisfies the Δ_2-condition in [86], and since the function $\frac{\psi(\lambda)}{\lambda^2}$ is non-increasing, we obtain that

$$l_\psi = \left\{\underline{a} \in \mathbb{R}^{\infty} : \sum_{k=1}^{\infty} \psi(|a_k|) < \infty\right\} \subseteq E_P.$$

where l_ψ is the Orlicz space. Since $\frac{\psi(\lambda)}{\lambda^2}$ is non-increasing, we also obtain that $l_\psi \subseteq l_2$.

Exercise 6.1.4. *Show that $E_p \subseteq l_2$.*

Let us make the following assumption:

(i) For $a \in E_P$, $ta \in E_P$, $0 \le t \le 1$.

Then assuming that

(ii) The function

$$H(t) = \sum_{k=1}^{\infty} \int_0^{\infty} (1-\cos(a_k u))\beta(u)\,du$$

is bounded on the interval $[0,1]$,

we can see that

$$\infty > \int_0^1 H(t)\,dt \;=\; \sum_{k=1}^{\infty} \int_{-\infty}^{\infty} \int_0^1 (1-\cos(ta_k u))\beta(u)\,dt\,du$$

$$=\; \sum_{k=1}^{\infty} \int_{-\infty}^{\infty} \left(1-\frac{\sin(a_k u)}{a_k u}\right)\beta(u)\,du.$$

But $1-\frac{\sin(x)}{x} \ge c\left(x^2 \wedge 1\right)$, giving that under assumptions (i) – (ii)

$$E_P = l_\psi.$$

We now prove that assumption (ii) is always satisfied. Note that

$$f(t) = \exp\left\{-\int_{-\infty}^{\infty}(1-\cos(tu))\beta(u)\,du\right\}$$

is a characteristic function of an infinitely divisible (compound Poisson) random variable X, and

$$g_n(t) = \exp\left\{-\sum_{k=1}^{n}\int_{-\infty}^{\infty}(1-\cos(a_k tu))\beta(u)\,du\right\}$$

is a characteristic function of a linear combination $\sum_{k=1}^{n} a_k X_k$ of n independent copies of X. Also,

$$\lim_{n\to\infty} g_n(t) = \exp\{-H(t)\} > 0$$

and for each t the limit is finite, giving that $\exp\{H(t)\}$ is a characteristic function of the infinite sum $\sum_{k=1}^{\infty} a_k X_k$. Hence, it is continuous and therefore bounded on a bounded interval, with the same conclusions following for the function H.

We recall that by general results on Orlicz spaces, $l_\psi = l_2$ if and only if there exist constants c_1, c_2, and λ_0, such that

$$0 < c_1 \le \frac{\psi(\lambda)}{\lambda^2} \le c_2 < \infty, \quad \lambda \le \lambda_0.$$

Hence, if $l_\psi = l_2$, we have

$$\lim_{\lambda\to 0} \int_{-\infty}^{\infty} \frac{\lambda^2 u^2 \wedge 1}{\lambda^2} \beta(u)\,du < \infty$$

giving that

$$\int_{-\infty}^{\infty} u^2 |\hat{h}(u)|^2\,du = \int_{-\infty}^{\infty} \left(\left(p^{1/2}(x)\right)'\right)^2 dx = \frac{1}{4}\int_{-\infty}^{\infty} \frac{(p'(x))^2}{p(x)}\,dx < \infty.$$

This proves that $E_P = l_2$ (in particular E_P is a subspace of l_2) if and only if

$$\int_{-\infty}^{\infty} \frac{(p'(x))^2}{p(x)}\,dx < \infty.$$

Thus, $E_P = l_2$ can hold for non-Gaussian measures, giving results in [27] and [13].

6.2 Equivalence and Singularity of Measures Generated by Gaussian Processes

Let $(\Omega, \mathcal{A})$ be a probability space, T an index set, and $I = \{i, i \text{ finite } \subseteq T\}$ be partially ordered under set inclusion. Consider a family of real random variables $\{X_t, t \in T\}$ defined on $(\Omega, \mathcal{A})$ and denote $\mathcal{F}_i = \sigma\{X_t, t \in i\}$ and $\mathcal{F} = \sigma\{X_t, t \in T\}$.

Suppose that P_1 and P_2 are two measures on $\mathcal{F}$, such that $\{X_t, t \in T\}$ is a Gaussian process on $(\Omega, \mathcal{F}, P_l)$, $l = 1, 2$. Assume that $E_{P_1} X_t = 0$ and $E_{P_2} X_t = m(t)$ and denote covariances $C_1(t,s) = E_{P_1} X_t X_s$ and $C_2(t,s) = E_{P_2}(X_t - m(t))(X_s - m(s))$.

Let $\mathcal{M}$ be the linear submanifold of the vector space of $\mathcal{F}$-measurable functions generated by the family $\{X_t, t \in T\}$ and $\mathcal{M}_i$ be the submanifold of $\mathcal{M}$ generated by $\{X_t, t \in i\}$ for each $i \in I$. We extend the functions C_l $(l = 1, 2)$ to $\mathcal{M}$ by defining

$$C_l(u,v) = \sum_{\substack{s \in i \\ t \in j}} a_s b_t C_l(s,t),$$

where $u = \sum_{s \in vi} a_s X_s$ and $v = \sum_{t \in vj} b_t X_t$ $(i, j \in I)$ for $l = 1, 2$. Hence, C_1, C_2 so extended are non-negative bilinear forms on $\mathcal{M}$ and, in particular, on $\mathcal{M}_i$ for each $i \in I$. For a fixed i, $\mathcal{M}_i$ is finite dimensional, which allows us to choose elements $u_{i1}, \ldots, u_{in_i} \in \mathcal{M}_i$, such that

$$C_1\left(u_{ij}, u_{ik}\right) = \delta_{jk} \quad \text{and} \quad C_2\left(u_{ij}, u_{ik}\right) = \delta_{jk}\lambda_{ik},$$

with $\lambda_{ik} > 0$ $(k = 1, \ldots, n_i)$ since the rank of C_1 equals the rank of C_2 on $\mathcal{M}_i$, as otherwise $P_1 \perp P_2$. Denote $m_{ik} = E_{P_2} u_{ik}$.

It is known that $P_1 \equiv P_2$ or $P_1 \perp P_2$ on $\mathcal{F}_i$. If $P_1 \equiv P_2$ on $\mathcal{F}_i$ we denote by ρ_i the Radon–Nikodym density of P_2 with respect to P_1 on $\mathcal{F}_i$ using the notation of Section 6.1. With the above notation we have the following lemma.

Lemma 6.2.1. *The following are equivalent:*

(a) P_1 *is not singular with respect to* P_2 *on* $\mathcal{F}$.

(b) $\sup_{i \in I} \sum_{k=1}^{n_i} \dfrac{\left(1 - \lambda_{ik}^{1/2}\right)^2}{2\lambda_{ik}^{1/2}} < \infty$ *and* $\sup_{i \in I} \sum_{k=1}^{n_i} \dfrac{m_{ik}^2}{\left(1 + \lambda_{ik}\right)^2} < \infty$.

(c) *There exists numbers* $0 < r_1 \leq r_2 < \infty$, *such that*

$$\begin{aligned}
&\text{(i)} \quad 0 < r_1 \leq \lambda_{ik} \leq r_2 \text{ for all } k = 1, \ldots, n_i \text{ and } i \in I, \\
&\text{(ii)} \quad \sup_{i \in I} \sum_{k=1}^{n_i} (1 - \lambda_{ik})^2 < \infty, \\
&\text{(iii)} \quad \sup_{i \in I} \sum_{k=1}^{n_i} m_{ik}^2 < \infty.
\end{aligned}$$

$$(6.4)$$

Proof. From (6.1) $P_1 \perp P_2$ on $\mathcal{F}$ if and only if $E_{P_1} \rho^{1/2} = 0$, where as in

Section 6.1, ρ is the Radon–Nikodym density of the absolutely continuous part of P_2 with respect to P_1. Hence, by Lemma 6.1.1, (a) is equivalent to $\inf_i E_{P_1} \rho_i^{1/2} > 0$. But

$$\rho_i = \left[\left(\prod_{k=1}^{n_i} \lambda_{ik} \right)^{-1} \right]^{1/2} \exp\left\{ -\frac{1}{2} \sum_{k=1}^{n_i} \left(\frac{(u_{ik} - m_{ik})^2}{\lambda_{ik}} - u_{ik}^2 \right) \right\}. \tag{6.5}$$

Hence, for $0 < \alpha < 1$,

$$E_{P_1} \rho_i^{\alpha} = \prod_{k=1}^{n_i} \left(\frac{\lambda_{ik}^{1-\alpha}}{\alpha + (1-\alpha)\lambda_{ik}} \right)^{1/2} \exp\left\{ -\frac{\alpha(1-\alpha)}{2} \sum_{k=1}^{n_i} \frac{m_{ik}^2}{(\alpha + (1-\alpha)\lambda_{ik})^2} \right\}. \tag{6.6}$$

Leaving (6.5) and (6.6) as an exercise, with $\alpha = 1/2$, we have

$$E_{P_1} \rho_i^{1/2} = \prod_{k=1}^{n_i} \left(\frac{2\lambda_{ik}^{1/2}}{1 + \lambda_{ik}} \right)^{1/2} \exp\left\{ -\frac{1}{4} \sum_{k=1}^{n_i} \frac{m_{ik}^2}{(1 + \lambda_{ik})^2} \right\}.$$

Hence (a) is equivalent to

$$\sup_{i \in I} \prod_{k=1}^{n_i} \left(\frac{1 + \lambda_{ik}}{2\lambda_{ik}^{1/2}} \right) < \infty \quad \text{and} \quad \sum_{k=1}^{n_i} \frac{m_{ik}^2}{(1 + \lambda_{ik})^2} < \infty. \tag{6.7}$$

Since $(1 + \lambda_{ik}) / \left(2\lambda_{ik}^{1/2} \right) = 1 + \left(1 - \lambda_{ik}^{1/2} \right)^2 / \left(2\lambda_{ik}^{1/2} \right)$ we obtain that condition (a) and (b) are equivalent.

Since $(1 + \lambda_{ik}) / \left(2\lambda_{ik}^{1/2} \right) > 1$ the first inequality in (6.7) implies the existence of real numbers $0 < r_1 \le r_2 < \infty$, such that condition (c)(i) is satisfied. Thus, (b) and (c)(i) are equivalent to (c). But (b) being equivalent to (6.7) implies (c)(i). The proof is complete. $\qquad \square$

Remark 6.2.1. *We note that condition (b) in Lemma 6.2.1 is equivalent to*

$$\sup_{i \in I} \sum_{k=1}^{n_i} \frac{(1 - \lambda_{ik})^2}{\lambda_{ik}} < \infty \quad \text{and} \quad \sup_{i \in I} \sum_{k=1}^{n_i} m_{ik}^2 < \infty. \tag{6.8}$$

Exercise 6.2.1. *Calculate the quantities in (6.5) and (6.6).*

Lemma 6.2.2. *Condition (6.4) implies that for every $\varepsilon > 0$, there exists $\alpha(\varepsilon) \in (0, 1)$, such that for all $i \in I$, $E_{P_1} \rho_i^{\alpha} > 1 - \varepsilon$ for $\alpha \in (0, \alpha(\varepsilon))$.*

Proof. Using the formulas (6.5) and (6.6) and taking the logarithm on both sides, we arrive at

$$-\ln E_{P_1} \rho_i^{\alpha} = \frac{1}{2} \left(\sum_{k=1}^{n_i} \ln\left(1 - \alpha\left(1 - \lambda_{ik}^{-1} \right) \right) - \alpha \ln\left(1 - \left(1 - \lambda_{ik}^{-1} \right) \right) \right)$$

$$+\frac{\alpha(1-\alpha)}{2}\sum_{k=1}^{n_i}\frac{m_i k^2}{(\alpha+(1-\alpha)\lambda_{ik})^2}.$$

Since for $-1 < a_1 < y < a_2 < \infty$ $(a_1 < 1 < a_2)$, we have [14]

$$y - dy^2 \le \ln(1-y) \le y - cy^2, \quad \text{with} \quad 0 < c < d, \tag{6.9}$$

using condition (6.4)(i) we have

$$-\ln E_{P_1}\rho_i^\alpha \le \alpha(d-c\alpha)\sum_{k=1}^{n_i}\left(1-\lambda_{ik}^{-1}\right)^2 + \alpha d_1 \sum_{k=1}^{n_i} m_{ik}^2, \tag{6.10}$$

for an appropriate constant d_1. We choose α_0 such that $(d-c\alpha) > 0$ for $\alpha \le \alpha_0$, then (6.10) and condition (6.4) imply that $\lim_{\alpha \to 0} \inf_{i \in I} E_{P_1}\rho_i^\alpha = 1$. □

By Theorem 6.1.1 we conclude that condition (6.4) implies $P_1 \ll P_2$ on $\mathcal{F}$. By the symmetry of the problem condition (6.4) implies $P_1 \equiv P_2$ on $\mathcal{F}$.

Theorem 6.2.1 (Dichotomy). *For two Gaussian measures P_1 and P_2 on $\mathcal{F}$, either $P_1 \perp P_2$ on $\mathcal{F}$ or $P_1 \equiv P_2$ on $\mathcal{F}$. Further, $P_1 \equiv P_2$ on $\mathcal{F}$ if and only if condition (6.4) holds.*

Corollary 6.2.1 (Hajek [42]). *Let $J_i = E_{P_1} = E_{P_1}(-\ln\rho_i) + E_{P_2}(\ln\rho_i)$. Then the following hold true:*

(a) $P_1 \perp P_2$ on $\mathcal{F}$ if and only if $\sup_i J_i = \infty$.

(b) $P_1 \equiv P_2$ on $\mathcal{F}$ if and only if $\sup_i J_i < \infty$.

Proof. From the form of ρ_i given in (6.5) and using the fact that the distribution of u_{ik} is $G(0,1)$ under P_1, we calculate

$$E_{P_1}\ln\rho_i = \frac{1}{2}\sum_{k=1}^{n_i}\left(\ln\frac{1}{\lambda_{ik}} - \frac{1}{\lambda_{ik}} + 1 - \frac{m_{ik}^2}{\lambda_{ik}}\right).$$

Similarly,

$$E_{P_2}\ln\rho_i = \frac{1}{2}\sum_{k=1}^{n_i}\left(-\ln\lambda_{ik} + \lambda_{ik} - 1 + m_{ki}^2\right).$$

Hence,

$$J_i = \frac{1}{2}\sum_{k=1}^{n_i}\left(\frac{(1-\lambda_{ik})^2}{\lambda_{ik}} + \frac{m_{ik}^2}{1+\lambda_{ik}}\right).$$

Thus, $\sup_{i \in I} J_i < \infty$ if and only if condition (6.4) holds in view of Lemma 6.2.1 and Remark 6.2.1. Now Theorem 6.2.1 implies the result. □

Since for every $i \in I$, the set $\{u_{i1},...,u_{in_i}\}$ generates the submanifold $\mathcal{M}_i$, then for every $t \in i$, $X_t = \sum_{k=1}^{n_i} a_{ik}(t)u_{ik}$. This implies that for $k = 1,...,n_i$,

$$a_{ik}(t) = C_1(X_t, u_{ik}) \quad \text{and} \quad \lambda_{ik}a_{ik}(t) = C_2(X_t, u_{ik}).$$

The elements $u_{i1}, ..., u_{ik}$ were selected to form an orthonormal basis in $\mathcal{M}_i$ as a subspace of $L_2(\Omega, \mathcal{F}_i, P_1)$, and the set

$$\left\{ \frac{u_{ik} - m_{ik}}{\lambda_{ik}^{1/2}}, k = 1, ..., n_i \right\}$$

is an orthonormal basis in $\mathcal{M}_i$ as a subspace of $L_2(\Omega, \mathcal{F}_i, P_2)$. Let $K(C_l^i)$ ($i \in I$) be the RKHS of C_l^i, the restriction of C_l to $i \times i$ ($l = 1, 2$). It follows by the above remarks that the set $\{a_{ik}, k = 1, ..., n_i\}$ is an orthonormal basis in $K(C_1^i)$ and $\left\{ \lambda_{ik}^{1/2} a_{ik}, k = 1, ..., n_i \right\}$ is an orthonormal basis in $K(C_2^i)$. Therefore, by Theorem 1.2.2, the set $\left\{ \lambda_{ik}^{1/2} a_{ik} a_{ij}, k, j = 1, ..., n_i \right\}$ is an orthonormal basis in $K\left(C_2^i \otimes C_1^i\right)$ and the set $\{a_{ik} a_{ij}, k, j = 1, ..., n_i\}$ is an orthonormal basis in $K\left(C_1^i \otimes C_1^i\right)$. For $t, s \in i$,

$$
\begin{aligned}
C_1^i(t,s) - C_2^i(t,s) &= \sum_{k=1}^{n_i} a_{ik}(t) a_{ik}(s) - \sum_{k=1}^{n_i} \lambda_{ik} a_{ik}(t) a_{ik}(s) \\
&= \sum_{k,j=1}^{n_i} \left(\lambda_{ik}^{-1/2} - \lambda_{ik} \right) \delta_{kj} \lambda_{ik}^{1/2} a_{ik}(t) a_{ij}(s).
\end{aligned}
$$

Hence,

$$C_1^i - C_2^i \in K\left(C_2^i \otimes C_1^i\right),$$

$$\left\| C_1^i - C_2^i \right\|_{K(C_2^i \otimes C_1^i)} = \sum_{k=1}^{n_i} \frac{(1 - \lambda_{ik})^2}{\lambda_{ik}} \quad \text{for} \quad i \in I. \tag{6.11}$$

Similarly,

$$C_1^i - C_2^i \in K\left(C_1^i \otimes C_1^i\right),$$

$$\left\| C_1^i - C_2^i \right\|_{K(C_1^i \otimes C_1^i)} = \sum_{k=1}^{n_i} (1 - \lambda_{ik})^2 \quad \text{for} \quad i \in I. \tag{6.12}$$

Furthermore, condition (6.4) on λ_{ik} implies that there exist constants $0 < \gamma_1 \le \gamma_2 < \infty$, such that for $u \in \mathcal{M}_i$ ($i \in I$)

$$\gamma_1 C_1(u,u) \le C_2(u,u) \le \gamma_2 C_1(u,u), \tag{6.13}$$

giving the following domination (recall Definition 1.2.2)

$$\gamma_1 C_1 \ll C_2 \ll \gamma_2 C_1. \tag{6.14}$$

Conversely, condition (6.14) implies (6.13). We now state a theorem relating the problem of equivalence of Gaussian measures to their covariances.

Theorem 6.2.2. *The following are equivalent:*

(a) $P_1 \equiv P_2$ *on* $\mathcal{F}$.

(b) $C_1 - C_2 \in K(C_2 \otimes C_1)$ *and* $m \in K(C_1)$.

(c) (i) *There exist constants* $0 < \gamma_1 \le \gamma_2 < \infty$, *such that* $\gamma_1 C_1 \ll C_2 \ll \gamma_2 C_1$;
    (ii) $C_1 - C_2 \in K(C_1 \otimes C_1)$ *and* $m \in K(C_1)$.

(d) $C_2 - C_1 \in K(C_1 \otimes C_2)$ *and* $m \in K(C_1)$.

(e) (i) *There exist constants* $0 < \delta_1 \le \delta_2 < \infty$, *such that* $\delta_1 C_2 \ll C_1 \ll \delta_2 C_2$;
    (ii) $C_2 - C_1 \in K(C_2 \otimes C_2)$ *and* $m \in K(C_1)$.

Violation of any condition implies that $P_1 \perp P_2$ *on* $\mathcal{F}$.

Proof. In view of the symmetry of the problem it suffices to prove the quivalence of (a), (b), and (c). To prove the equivalence of (a) and (c) we observe that from (6.12), Theorem 6.2.1, (6.4), and (6.13), we conclude that (a) is equivalent to (6.14) and

$$\sup_{i \in I} \left\| C_1^i - C_2^i \right\|_{K(C_1^i \otimes C_2^i)} < \infty \quad \text{and} \quad \sup_{i \in I} \left\| m_i \right\|_{K(C_1^i)} = \sup_{i \in I} \left\| \sum_{k=1}^{n_i} m_{ik} a_{ik} \right\|_{K(C_1)} < \infty,$$

where $m_i(t)$ denotes restriction of $m(t)$ to $t \in i$. Now, condition (6.14) is equivalent to (6.4)(i) and the remaining two conditions are equivalent to (c)(ii), giving that condition (a) is equivalent to (c) by Exercise 1.2.8. By (6.11) and the first part of the proof, condition (b) is equivalent to (6.8). By Lemma 6.2.1 and Remark 6.2.1, conditions (b) and (a) are equivalent. $\square$

Let $(\Omega, \mathcal{A})$ be a probability space, $\{X_t, t \in T\}$ a family of real random variables, I, $\mathcal{F}_i$ and $\mathcal{F}$ be as before, and P_1, P_2 be two measures on $\mathcal{F}$ such that $\{X_t, t \in T\}$ is a Gaussian process on $(\Omega, \mathcal{F}, P_l)$ $(l = 1, 2)$, with $E_{P_l} X_t = m_l(t)$ and covariances $C_l(t, s) = E_{P_l} (X_t - m_l(t))(X_s - m_l(s))$, $l = 1, 2$. Then $P_1 \equiv P_2$ if and only if $P_1' \equiv P_2'$, where P_1' is the measure induced on $\mathcal{F}$ by the process $\{X_t - m_1(t), t \in T\}$ under P_l $(l = 1, 2)$. Since $E_{P_1'} X_t = 0$, $E_{P_2'} X_t = m_2(t) - m_1(t)$, and $C_l(t, s) = E_{P_l'} (X_t - m_l(t))(X_s - m_l(s))$ we obtain the following theorem by applying Theorem 6.2.2.

Theorem 6.2.3. *The following are equivalent:*

(a) $P_1 \equiv P_2$ *on* $\mathcal{F}$.

(b) $C_1 - C_2 \in K(C_2 \otimes C_1)$ *and* $m_2 - m_1 \in K(C_1)$.

(c) (i) *There exist constants* $0 < \gamma_1 \le \gamma_2 < \infty$, *such that* $\gamma_1 C_1 \ll C_2 \ll \gamma_2 C_1$;
    (ii) $C_1 - C_2 \in K(C_1 \otimes C_1)$ *and* $m - 2 - m_1 \in K(C_1)$.

(d) $C_2 - C_1 \in K(C_1 \otimes C_2)$ *and* $m_1 - m_2 \in K(C_1)$.

(e) (i) *There exist constants* $0 < \delta_1 \le \delta_2 < \infty$, *such that* $\delta_1 C_2 \ll C_1 \ll \delta_2 C_2$;
    (ii) $C_2 - C_1 \in K(C_2 \otimes C_2)$ *and* $m_1 - m_2 \in K(C_1)$.

Violation of any condition implies that $P_1 \perp P_2$ on $\mathcal{F}$.

Equivalence (a) $\Leftrightarrow$ (b) of Theorem 6.2.3 is due to Parzen [103] (see also Neveu [92]).

Using Theorem 6.2.3 (c)(i) and Exercise 1.2.7 we can obtain a non-negative definite linear bounded operator $L \in \mathcal{L}(K(C_1))$ defined by $(Lf)(t) = \langle f, C_2(\cdot,t) \rangle_{K(C_1)}$. Hence

$$((I-L)f)(t) = \langle f, C_1(\cdot,t) - C_2(\cdot,t) \rangle_{K(C_1)}.$$

Here I denotes the identity operator on $K(C_1)$. Using Theorem 6.2.3 (c) and Lemma 1.2.1 we obtain the following result.

Theorem 6.2.4. *The following are equivalent:*

(a) $P_1 \equiv P_2$ *on* $\mathcal{F}$.

(b) (i) *There exist constants* $0 < \gamma_1 \leq \gamma_2 < \infty$, *such that* $\gamma_1 C_1 \ll C_2 \ll \gamma_2 C_1$;

    (ii) $(I-L) \in \mathcal{L}_2(K(C_1))$ *(Hilbert–Schmidt operators on* $(K(C_1))$*)*;

    (iii) $m_2 - m_1 \in K(C_1)$.

(c) (i) *The operator* $L \in \mathcal{L}(K(C_1))$ *given by* $(Lf)(t) = \langle f, C_1(\cdot,t) \rangle_{K(C_1)}$, $f \in K(C_1)$, *is such that* $(I-L) \in \mathcal{L}_2(K(C_1))$;

    (ii) 1 *is not an eigenvalue of* $(I-L)$;

    (iii) $m_2 - m_1 \in K(C_1)$.

The equivalence of (a) and (b) follows from the fact that in the case $(I-L)$ being Hilbert–Schmidt, one is not an eigenvalue of $(I-L)$ if and only if L is invertible.

Remark 6.2.2. *The condition (a) (iii) can be restated by saying that L has a pure point spectrum and $\sum_n (1-\lambda_n)^2 < \infty$ for nonzero eigenvalues λ_n of L. In this form, Theorem 6.2.4 was proved by Kallianpur and Oodaira [55] (see also [26]).*

6.3 Conditions for Equivalence: Special Cases

In this section, $P = G(m,C)$ will mean that the process $\{X_t, t \in T\}$ is Gaussian under P with mean $m(t) = E_P X_t$ and covariance $C(s,t) = E_P(X_t - m(t))(X_s - m(s))$.

6.3.1 Introduction

We begin with two fundamental examples of processes equivalent to Wiener–Lévy Brownian Motion and Cameron–Yeh process.

1. Gaussian Process Equivalent to Wiener–Lévy Brownian Motion [50], [115]. Let P_1 be $G(0,C_1)$ with $C_1(t,s) = \min(t,s) = \int_0^\infty 1_{(0,t]}(u)1_{(0,s]}(u)\,du$, $(s,t \in \mathbb{R}_+)$. Then $P_2 = G(m,C)$ is equivalent to P_1 if and only if

(a) $m(t) = \int_0^t f(u)\,du$ for some $f \in L^2(\mathbb{R}_+)$.

(b) $C(t,s) = \min(t,s) - \int_0^t \int_0^s g(u,v)\,du\,dv$ for some symmetric $g \in L^2(\mathbb{R}_+^2)$.

(c) 1 is not an eigenvalue of the integral operator

$$J(h)(u) = \int_0^\infty g(u,v)h(v)\,dv.$$

Exercise 6.3.1. *Prove the statement above.*
Hint: Use Theorem 6.2.1 and Theorem 2.3.1. Note that

$$(C_1 \otimes C_1)((t_1,t_2),(s_1,s_2))$$

$$\int_0^\infty \int_0^\infty 1_{(0,t_1] \times (0,s_1]}(u,v) 1_{(0,t_2] \times (0,s_2]}(u,v)\,du\,dv.$$

2. Gaussian Process Equivalent to Cameron-Yeh process [102]. Let $T = \mathbb{R}_+^n$ and $P_1 = G(0,C_1)$ with $C_1(t,s) = \prod_{k=1}^n \min(t_k,s_k)$ for $t = (t_1,...,t_n)$, $s = (s_1,...,s_n)$ Then $P_2 = G(m,C)$ is equivalent to P_1 if and only if

(a) $m(t) = \int_T R_t(u)f(u)\,du$ for some $f \in L^2(T)$, where

$$R_t(u) = 1_{(0,t_1] \times ... \times (0,t_n]}(u_1,...,u_n).$$

(b) $C(t,s) = C_1(t,s) - \int_T \int_T R_t(u)R_t(s)g(u,v)\,du\,dv$ for some symmetric $g \in L^2(T \times T)$.

(c) 1 is not an eigenvalue of the integral operator

$$J(h)(u) = \int_T g(u,v)h(v)\,dv$$

on $L_2(T)$.

Exercise 6.3.2. *Prove the statement above.*
Hint: Use Theorem 6.2.2, Theorem 6.2.4, and Theorem 2.3.1, noting the form of the covariance $C_1 \otimes C_1$.

6.3.2   Gaussian Processes with Independent Increments

Following the discussion in [117], we consider a measurable space $(\mathcal{X},\mathcal{A})$ and a Gaussian process $\{X_t, t \in T\}$ defined on a probability space $\{\Omega,\mathcal{F},P\}$ with the index set $T = \{\mathcal{A}\}$. The process X_t is said to have *independent increments* if there exists a σ-finite measure μ on $\mathcal{A}$ such that $E_p X_t = m(t)$ and $EX_t X_s = \mu(t \cap s)$ for $t,s \in \{A, A \in \mathcal{A}, \mu(A) < \infty\}$.

Theorem 6.3.1. *Let $P_1 = G(0,\mu_1)$ and $P_2 = G(m,\mu_2)$, where μ_1,μ_2 are σ-finite measures μ on $\mathcal{A}$, then $P_1 \equiv P_2$ if and only if*

(a) $\mu_1^{(c)} = \mu_2^{(c)}$, *where $\mu_i^{(c)}$ denotes the nonatomic part of μ_i (i = 1,2),*

(b) μ_1 and μ_2 have the same atoms $\{a_n, n = 1, 2, ...\}$, $a_n \in \mathcal{X}$, and

$$\sum_{n=1}^{\infty} \left(1 - \frac{\mu_2(\{a_n\})}{\mu_1(\{a_n\})}\right)^2 < \infty,$$

(c) $m \ll \mu_1$ with $(dm/d\mu_1) \in L^2(\mathcal{X}, \mathcal{A}, \mu_1)$.

Proof. We first note that since $C_1(t,s) = \int_{\mathcal{X}} 1_t(u) 1_s(u) d\mu_1(u)$, it follows from Theorem 6.2.2 that $P_1 \equiv P_2$ if and only if

(a') $\mu_2(t \cap s) = \mu_1(t \cap s) - \iint_{\mathcal{X} \times \mathcal{X}} 1_t(u) 1_s(v) g(u,v) d\mu_1(u) d\mu_2(v)$ for some $g \in L^2(\mathcal{X} \times \mathcal{X}, \mathcal{A} \otimes \mathcal{A}, \mu_1 \otimes \mu_2)$.

(c') $m(t) = \int_{\mathcal{X}} 1_t(u) f(u) d\mu_1(u)$ for some $f \in L^2(\mathcal{X}, \mathcal{A}, \mu_1)$.

Here, we have used Example 1.2.1(f), and the fact that, by Exercise 1.2.2 and Theorem 1.2.2, the tensor product of covariances can be written as

$$(C_1 \otimes C_2)[(t_1,t_2),(s_1,s_2)] = \iint_{\mathcal{X} \otimes \mathcal{X}} 1_{t_1 \times s_1}(u,v) 1_{t_2 \times s_2}(u,v) d\mu_1(u) d\mu_2(v).$$

Now, (c) is equivalent to (c') above. In view of (a'), the signed measure $\int_t \int_s g(u,v) d\mu_1 u d\mu_2(v)$ vanishes off the diagonal as a function of $t \cap s$. If μ_1 is nonatomic, then we get from (a') that $\mu_1(t \cap s) = \mu_2(t \cap s)$, that is, $\mu_1 = \mu_2$. In the case when there exists $a \in \mathcal{X}$, such that $\mu_1(\{a\}) \neq 0$ and $\mu_2(\{a\}) = 0$, again we obtain from (a') that $\mu_1 = \mu_2$ as the integral term vanishes, giving a contradiction.

Hence, μ_1 and μ_2 have positive measures at the same set of atoms in $\mathcal{X}$. Since μ_1 and μ_2 are σ-finite, there exists at most countable set of atoms $\{a_n, n = 1, 2, ...\} \subseteq \mathcal{X}$ with $\mu_1(\{a_n\}) = \mu_2(\{a_n\})$. In this case, condition (a') is equivalent to (a) $\mu_1^{(c)} = \mu_2^{(c)}$ and (b') $\mu_2(\{a_n\}) = \mu_1(\{a_n\}) - g(a_n,a_n)\mu_1(\{a_n\})\mu_2(\{a_n\})$. Now, condition (b') is equivalent to (b) since the series $\sum_{n=1}^{\infty} g^2(a_2,a_n)\mu_1(\{a_n\})\mu_2(\{a_n\})$ converges by the assumption that $g \in L^2(\mathcal{X} \times \mathcal{X}, \mathcal{A} \otimes \mathcal{A}, \mu_1 \otimes \mu_2)$. $\square$

6.3.3 Stationary Gaussian Processes

Following [26] and [117] we formulate an equivalence theorem for measures related to stationary Gaussian processes.

Theorem 6.3.2. *Let T be a locally compact abelian group with a separable dual $\hat{T}$ and operation $+$. Let $P_1 = G(0, C_1)$ and $P_2 = G(0, C_2)$, where $C_i(t,s) = R_i(t-s)$, and $R_i(t)$ is a continuous nonnegative definite function on T. Denote by μ_1 and μ_2 the associated spectral measures. Then $P_1 \equiv P_2$ if and only if*

(a) $\mu_1^{(c)} = \mu_2^{(c)}$, *where $\mu_i^{(c)}$ denotes the nonatomic part of μ_i $(i = 1, 2)$,*

(b) μ_1 *and μ_2 have the same atoms $\{a_n, n = 1, 2, ...\}$, $a_n \in \mathcal{X}$, and*

$$\sum_{n=1}^{\infty} \left(1 - \frac{\mu_2(\{a_n\})}{\mu_1(\{a_n\})}\right)^2 < \infty.$$

Proof. It follows from Exercise 6.3.3 and Theorem 6.2.2 that $P_1 \equiv P_2$ if and only if

$$R_2(t-s) = R_1(t-s) - \iint_{\hat{T} \times \hat{T}} \langle t, u \rangle \langle s, v \rangle g(u, v) d\mu_1(u) d\mu_2(v),$$

with $g \in L^2(\hat{T} \times \hat{T}, \mu_1 \otimes \mu_2)$. Since the last term depends on $t - s$, the signed measure $\iint_{A \times B} g(u, v) d\mu_1(u) d\mu_2(v)$ $(A \times B \subseteq \hat{T} \times \hat{T})$ is zero off the diagonal by Bochner's theorem and the uniqueness of Fourier transform. An argument as in the proof of Theorem 6.3.1 shows that μ_1 and μ_2 have the same atoms $\{a_n, n = 1, 2, ...\} \subseteq \hat{T}$. Hence, we have

$$R_2(t) = R_1(t) - \sum_{n=1}^{\infty} e^{ita_n} g(a_n, a_n) d\mu_1(\{a_n\}) d\mu_2(\{a_n\}).$$

Using the uniqueness of Fourier transform we conclude that

$$\mu_2(A) = \mu_1(A) - \sum_{n=1}^{\infty} g(a_n, a_n) \mu_1(\{a_n\}) \mu_2(\{a_n\}).$$

Now the proof is completed as for Theorem 6.3.1. □

Exercise 6.3.3. *Let T be a locally compact abelian group with a separable dual $\hat{T}$ and operation $+$. Consider covariance $C(t, s) = R(t - s)$, where $R(t)$ is a continuous nonnegative definite function on T. It is known, see [112], that the function $R(t)$ is given by*

$$R(t) = Re\left(\int_{\hat{T}} \langle t, u \rangle \, d\mu(u) \right),$$

where μ is a nonnegative finite measure on the dual $\hat{T}$, and $\langle t, u \rangle$ denote the duality between T and $\hat{T}$. Show that

$$R(t - s) = Re\left(\int_{\hat{T}} \langle t, u \rangle \langle s, u \rangle \, d\mu(u) \right)$$

and

$$K(C) = \left\{ h: h(t) = Re\left(\int_{\hat{T}} t(u) g(u) \, d\mu(u) \right) \right\},$$

where $g \in L^2(\hat{T}, \mu)$ and

$$\langle h_1, h_h \rangle_{K(C)} = \int_{\hat{T}} g_1(u) g_2(u) \, d\mu(u).$$

Show that

$$K(C \otimes C) = \left\{ h: h(t, s) = \iint_{\hat{T} \times \hat{T}} \langle t, u \rangle \langle s, v \rangle g(u, v) \, d\mu(u) d\mu(v) \right\},$$

where $g \in L^2(\hat{T} \times \hat{T}, \mu \otimes \mu)$.

6.3.4   Gaussian Measures on Banach Spaces

Let E be a real separable Banach space and $\mathcal{B}(E)$ the Borel subsets of E. For a centered Gaussian measure P on E, let $H(P)$ denote the subspace of $L^2(E, \mathcal{B}(E), P)$ generated by all continuous linear functionals on E, $\{\langle x^*, \cdot \rangle, x^* \in E^*\}$, with $\langle \cdot, \cdot \rangle$ denoting the duality on $E^* \times E$. We denote by $C_P : E^* \to E$ the operator defined by

$$C_P(x^*) = \int_E \langle x^*, x \rangle x \, dP(x),$$

where the integrals are in the sense of Bochner [17]. We note that for $x^*, y^* \in E^*$, the function

$$C_P(x^*, y^*) = \langle y^*, C_P x^* \rangle$$

is the covariance of the Gaussian process $\{\langle x^*, \cdot \rangle, x^* \in E^*\}$. Let us denote the measure P by $G(0, C_P)$. Let $Q = G(0, C_Q)$ with mean $m = \int_E x \, dQ(x) \in E$ and the corresponding operator $C_Q(x^*) = \int_E \langle x^*, x - m \rangle (x - m) \, dQ(x)$. We have the following result.

Theorem 6.3.3. *With the above notation, $P \equiv Q$ if and only if*

(a) $m = \int_E x g_0(x) \, dP(x)$ *for some $g_0 \in H(P)$,*

(b) $C_P - C_Q = G_0$, *where $G_0 : E^* \to E$ is an operator defined by*

$$G_0(x^*) = \iint_{E \times E} f_0(x, y) \langle x^*, x \rangle y \, dP(x) dQ(y),$$

with the function f_0 being symmetric and an element of the linear subspace of $L^2(E \times E, \mathcal{B}(E) \otimes \mathcal{B}(E), P \otimes Q)$ generated by the set $\{\langle x^*, \cdot \rangle \langle y^*, \cdot \rangle, x^*, y^* \in E^*\}$.

Proof. Theorem 6.2.2 and Example 5.1.1 imply that $P \equiv Q$ if and only if

(a') $\langle x^*, m \rangle = \int_E \langle x^*, x \rangle g(x) \, dP(x)$,

(b') $C_P(x^*, y^*) = C_Q(X^*, y^*) - \iint_{E \times E} f_0(x, y) \langle x^*, x \rangle \langle y^*, y \rangle \, dP(x) dQ(y)$.

The new conditions (a') and (b') are equivalent to (a) and (b).   □

Using Theorem 6.2.4 we can restate Theorem 6.3.3 as follows.

Theorem 6.3.4. *With the above notation, $P \equiv Q$ if and only if*

(a) $m = \int_E x g_0(x) \, dP(x)$ *for some $g_0 \in H(P)$,*

(b) $C_P - C_Q = \tilde{G}$, *where $\tilde{G} : E^* \to E$ is an operator defined by*

$$\tilde{G}(x^*) = \iint_{E \times E} \tilde{g}(x, y) \langle x^*, x \rangle y \, dP(x) dP(y),$$

with the function $\tilde{g}$ being symmetric and an element of the linear subspace of $L^2(E \times E, \mathcal{B}(E) \otimes \mathcal{B}(E), P \otimes P)$ generated by the set $\{\langle x^*, \cdot \rangle \langle y^*, \cdot \rangle, x^*, y^* \in E^*\}$,

(c) *The operator* $L : H(P) \to H(P)$ *defined by*

$$Lf = \int_E \tilde{g}(x,y) f(y) \, dP(x)$$

does has not have unity as an eigenvalue.

With $E = C([0,1])$, the Banach space of continuous functions on $[0,T]$ with the supremum norm, and restricting to $x^* = \varepsilon_t$, the unit mass at t, the condition in Theorem 6.3.4 gives the result of Shepp [115].

Let us now consider $E = l_p$ $(p \geq 1)$, the Banach space of real sequences summable with power p, see [65]. Let

$$C_P = \left(s_{ij}^P \right)_{i,j=1}^{\infty}, \quad C_Q = \left(s_{ij}^Q \right)_{i,j=1}^{\infty},$$

be matrices of covariance operators on measures $P = G(0, C_P)$ and $Q = G(m, C_q)$, and assume that the matrix C_P is diagonal. Let $e_i = (0, ..., 0, \underset{i}{1}, 0, ...)$. We note that

$$H(P) = \left\{ \sum_{i=1}^{\infty} a_i e_i(\cdot) : \sum_{i=1}^{\infty} a_i^2 s_{ii}^P < \infty, \ \{a_i\}_{i=1}^{\infty} \subseteq \mathbb{R} \right\} \subseteq L^2(E, \mathcal{B}(E), P),$$

and the subspace of $L^2(E \times E, \mathcal{B}(E) \otimes \mathcal{B}(E), P \otimes P)$ where lies the element f_0 of condition (b) in Theorem 6.3.3 is

$$\left\{ \sum_{i,j=1}^{\infty} b_{i,j} e_i(\cdot) e_j(\cdot) : \sum_{i,j=1}^{\infty} b_{i,j}^2 s_{ii}^P s_{jj}^P < \infty, \{b_{ij}\}_{i,j=1}^{\infty} \subseteq \mathbb{R} \right\}.$$

Hence, we have the following result.

Theorem 6.3.5. *With the previous notation,* $P \equiv Q$ *if and only if*

(a) $e_i(m) = a_i S_{ii}^P$ *for all* $i \geq 1$ *with* $\sum_{i=1}^{\infty} a_i^2 s_{ii}^P < \infty$,

(b) $s_{ij}^Q = s_{ij}^P - b_{ij} s_{ii}^P s_{jj}^P$, *for some real numbers* b_{ij} *with* $\sum_{i,j=1}^{\infty} b_{i,j}^2 s_{ii}^P s_{jj}^P < \infty$.

(c) *The operator* $B : l^p \to l^p$ *given by the matrix* $\left(b_{ij} \right)_{i,j=1}^{\infty}$ *has no eigenvalue equal to* 1.

6.3.5 Generalized Gaussian Processes Equivalent to Gaussian White Noise of Order p

Following [47] we consider an index set $T = C_0^{\infty}(V)$, where V is an open subset of $\mathbb{R}^n$ and covariance $C_P(t,s) = \sum_{|\alpha| \leq p} \int_V D^\alpha t(u) D^\alpha s(u) \, du$, similar as in Example 1.6 (d). Let $P = G(0, C)$ and $Q = G(m, C_0)$, where m is a continuous function on T and $C_Q(\cdot, \cdot)$ is a continuous bilinear form on $T \times T$ under Schwartz's topology. The following theorem is a direct consequence of Theorem 6.2.4 and Examples 1.6 and 2.2.1 (e).

Theorem 6.3.6. *With the previous notation, $P \equiv Q$ if and only if*

(a) $m = \int_V f(u)t(u)\,du$ *for some* $f \in W_0^{p,2}(V)$,

(b) $C_P(t,s) - C_Q(t,s) = \iint_{V \times V} g_0(u,v)t(u)s(u)\,du\,dv$, *for some symmetric $g_0 \in W_0^{p,2}(V \times V)$.*

(c) *The operator $G_0 : W_0^{p,2}(V) \to W_0^{p,2}(V)$ defined by*

$$(G_0 f)(v) = \int_V g_0(u,v)f(v)\,du$$

does has not have unity as an eigenvalue.

6.4 Prediction or Kriging

We now consider some applications to spatial statistics [120]. Suppose we have spatial data collected in a domain D. As an example we can consider meteorological data at different weather stations.

Let $\{X_t, t \in D\}$ be measurements obtained at locations $\{t_k, k = 1,2...\}$. Assume that the set of sampling is dense. Let $\hat{X}_i(t,n)$ $(i = 0,1)$ be prediction of $X(t)$ based on observations made at locations $(t_1,...,t_n)$ under two probability measures P_0 and P_1, that is,

$$\hat{X}_i(t,n) = E_{P_i}\left(X_t \,\middle|\, X_{t_1},...,X_{t_n}\right), \quad i = 0,1.$$

Denote the prediction errors by $e_i(t,n) = X_t - \hat{X}_i(t,n)$ $(i = 0,1)$.

In general, if $h_1,...,h_n$ are observation points, we consider vectors $\psi_1,...,\psi_n$ obtained from the observations using Gram–Schmidt orthogonalization under P_0, that is, $E_{P_0} \psi_i \psi_j = \delta_{ij}$.

Let $\mathcal{M}_k$ denote the submanifold of measurable functions generated by X_{t_j}, $j \leq k$. For $i = 1,2$, denote by $H_i(X)$ the completion of $\bigcup_k \mathcal{M}_k$ in $L^2(\Omega,\mathcal{F},P_i)$. Clearly, if $P_0 \equiv P_1$, then $H_0(X) = H_1(X)$. Define a map $\Lambda : H_0(X) \to H_1(X)$, $\Lambda u = u$, then from the general theorem on equivalence and singularity, Theorem 6.2.2, for some $0 < \gamma_1 \leq \gamma_2 < \infty$,

$$\gamma_1 \|u\|_{H_0(X)}^2 \leq \|\Lambda u\|_{H_1(X)}^2 \leq \gamma_2 \|u\|_{H_0(X)}^2,$$

giving that $\Lambda : H_0(X) \to H_1(X)$ is a bounded operator with bounded inverse. Let $\pi_i : K(C_i) \to H_i(X)$ $(i = 0,1)$ be the canonical isometries and

$$\tilde{\Lambda} = \pi_1^* \Lambda \pi_0, \quad \left(\pi_i^* = \pi_i^{-1}, i = 0,1\right).$$

Then $\tilde{\Lambda} : K(C_0) \to K(C_1)$ and

$$\tilde{\Lambda}^* \tilde{\Lambda} = \pi_0^* \Lambda^* \Lambda \pi_0 : K(C_0) \to K(C_0)$$

and

$$\langle \pi_0^* \Lambda^* \Lambda \pi_0 C_0(\cdot,t), C_0(\cdot,s) \rangle_{K(C_0)} = \langle \Lambda X_t, \Lambda X_s \rangle_{H_1(X)} = C_1(t,s),$$

showing that

$$\pi_0^* \Lambda^* \Lambda \pi_0 C_0(\cdot, t) = L(C_0(\cdot, t)), \quad t \in D,$$

where, as in Section 6.2, the operator $L : K(C_0) \to (C_0)$ is defined by $(Lf)(t) = \langle f, C_1(\cdot, t) \rangle_{K(C_0)}$. Hence

$$(I_0 - L) = \pi_0^* (I_0 - \Lambda^* \Lambda) \pi_0,$$

where I_0 denotes identity on $H_0(X)$. We have the following result.

Theorem 6.4.1. *With the previous notation, $P_0 \equiv P_1$ if and only if*

(i) $\Lambda : H_0(X) \to H_1(X)$ *is a one-to-one bounded operator with bounded inverse,*

(ii) $(I_0 - \Lambda^* \Lambda) : H_0(X) \to H_0(X)$ *is a Hilbert–Schmidt operator,*

(iii) $m_1 - m_0 \in K_{(C_0)}.$

The next theorem is proved in [120].

Theorem 6.4.2. *With the notation of this section, let*

$$\mathcal{H}_{-n} = \left\{ h \in H_0(X) : E_0 e_0^2(h, n) > 0 \right\}.$$

If $P_0 \equiv P_1$, then

$$\lim_{n \to \infty} \sup_{\psi \in \mathcal{H}_{-n}} \left| \frac{E_{P_1} e_0^2(\psi, n) - E_{P_0} e_0^2(\psi, n)}{E_{P_0} e_0^2(\psi, n)} \right| = 0$$

$$\lim_{n \to \infty} \sup_{\psi \in \mathcal{H}_{-n}} \left| \frac{E_{P_0} e_1^2(\psi, n) - E_{P_1} e_1^2(\psi, n)}{E_{P_1} e_1^2(\psi, n)} \right| = 0$$

$$\lim_{n \to \infty} \sup_{\psi \in \mathcal{H}_{-n}} \left| \frac{E_{P_0} e_1^2(\psi, n) - E_{P_0} e_0^2(\psi, n)}{E_{P_0} e_0^2(\psi, n)} \right| = 0$$

Proof. If $P_0 \equiv P_1$, then $(I_0 - \Lambda^* \Lambda)$ is a Hilbert–Schmidt operator on $H_0(X)$, so that

$$\sum_{j=1}^{\infty} \left\| (I_0 - \Lambda^* \Lambda) \psi_j \right\|_{H_0(X)}^2 < \infty \quad \text{and} \quad \sum_{j=1}^{\infty} m_{ij}^2 < \infty,$$

where $m_{1j} = E_{P_1} (\psi_j)$. Any $\psi \in H_0(X)$ can be expanded as $\psi = \sum_{j=1}^{\infty} c_j \psi_j$, where $\sum_{j=1}^{\infty} c_j^2 < \infty$. Then the error of linear prediction for ψ given $H_0(X)$ is

$$e_0(\psi, n) = \sum_{j=n+1}^{\infty} c_j \psi_j \quad \text{and} \quad E_{P_0} e_0^2(\psi, n) > 0.$$

Now,

$$\left| \frac{E_{P_1} e_0^2(\psi, n) - E_{P_0} e_0^2(\psi, n)}{E_{P_0} e_0^2(\psi, n)} \right|$$

$$= \left| \frac{E_{P_1}(\Lambda e_0(\psi,n))^2 - E_{P_0}e_0^2(\psi,n)}{E_{P_0}e_0^2(\psi,n)} \right|$$

$$= \left| \langle \Lambda^* \Lambda e_0(\psi,n), e_0(\psi,n) \rangle_{H_0(X:n)} - \langle e_0(\psi,n), e_0(\psi,n) \rangle_{H_0(X)} \right.$$

$$\left. + \left(\sum_{j=n+1}^{\infty} c_j m_{0j} \right)^2 \right| / \sum_{j=n+1}^{\infty} c_j^2$$

$$\leq \frac{\left| \langle (I_0 - \Lambda^* \Lambda) e_0(\psi,n), e_0(\psi,n) \rangle_{H_0(X)} + \sum_{j=n+1}^{\infty} c_j^2 \sum_{j=n+1}^{\infty} m_{0j}^2 \right|}{\sum_{j=n+1}^{\infty} c_j^2}$$

$$\leq \frac{\sum_{j=1}^{\infty} |c_j| \, \| (I_0 - \Lambda^* \Lambda) \psi_j \|_{H_0(X)} \| e_0(\psi,n) \|_{H_0(X)}^2 + \sum_{j=n+1}^{\infty} c_j^2 \sum_{j=n+1}^{\infty} m_{0j}^2}{\sum_{j=n+1}^{\infty} c_j^2}$$

Now, since $\| e_0(\psi,n) \|_{H_0(X)} = \left(\sum_{j=n+1}^{\infty} c_j^2 \right)^{1/2}$, then by the Schwartz inequality the numerator in the last expression is dominated by

$$\left(\sum_{j=1}^{\infty} \| (I_0 - \Lambda^* \Lambda) \psi_j \|_{H_0(X)}^2 \right)^{1/2} + \sum_{j=n+1}^{\infty} m_{0j}^2,$$

which converges to zero as $n \to \infty$ independent of ψ. The existence of the other two limits follows by a similar argument by interchanging indexes 0 and 1. $\square$

Following the result of Theorem 6.4.2, if $E_{P_0}e_1^2(\psi,n) \to 0$, we call the predictor $\hat{X}_1(t,n)$ *asymptotically optimal* under P_0. Theorem 6.4.2 shows that if $P_1 \equiv P_0$, then $\hat{X}_1(t,n)$ is asymptotically optimal under P_0.

We now consider the case of a centered stationary Gaussian random field.

Theorem 6.4.3. *Let, under the probability measures P_j ($j = 0, 1$), $\{X_t, \, t \in D\}$ be a centered stationary Gaussian random field with covariance C_j and spectral measure F_j with density f_j. The index set $D \subseteq \mathcal{R}^d$ is assumed bounded. Let $b(s,t) = C_0(s,t) - C_1(s,t)$, $s,t \in D$, where*

$$C_j(s,t) = \int_{\mathbb{R}^d} e^{i\lambda(s-t)} f_j(\lambda) \, d\lambda.$$

Suppose there exists constants $0 < \gamma_1 \leq \gamma_2 < \infty$, such that $\gamma_1 C_0 \ll C_1 \ll \gamma_2 C_0$.

A necessary and sufficient condition for the equivalence of probability measure P_0 and P_1 is that the function $b(s,t)$, $s,t \in D$ can be extended to a square–integrable function on $\mathbb{R}^d \times \mathbb{R}^d$ whose Fourier transform $\psi(\lambda,\mu)$ satisfies

$$\iint_{\mathbb{R}^d \times \mathbb{R}^d} \frac{|\psi(\lambda,\mu)|^2}{f_0(\lambda) f_0(\mu)} \, d\lambda \, d\mu < \infty.$$

Proof. It follows from Theorem 6.2.2 (c) that, under our assumptions, $P_0 \equiv P_1$ if and only if $b(s,t) \in K(C_0 \otimes C_0)$, an RKHS consisting of functions $\hat{\phi}$ of the following form:

$$\hat{\phi}(s,t) = \iint_{\mathbb{R}^d \times \mathbb{R}^d} e^{-i\left(\langle s,\lambda\rangle_{\mathbb{R}^d}\langle t,\mu\rangle_{\mathbb{R}^d}\right)} \varphi(\lambda,\mu)\, dF_0(\lambda)\, dF_0(\mu),$$

with $\varphi \in L^2(F_0 \otimes F_0)$.

Also, for any $\hat{\phi} \in K(C_0 \otimes C_0)$ there exists a function $\varphi \in L^2(F_0 \otimes F_0)$ such that

$$\hat{\phi}(s,t) = \langle \hat{\phi}(\cdot), C_0 \otimes C_0((s,t),\cdot)\rangle_{K(C_0 \otimes C_0)}$$
$$= \iiiint_{\mathbb{R}^d \times \mathbb{R}^d \times \mathbb{R}^d \times \mathbb{R}^d} e^{-i\left(\langle s,\lambda\rangle_{\mathbb{R}^d}\langle t,\mu\rangle_{\mathbb{R}^d}\right)} e^{-i\langle t-t_1,\lambda\rangle_{\mathbb{R}^d} - i\langle s-s_1,\mu\rangle_{\mathbb{R}^d}}$$
$$\times \varphi(\lambda,\mu)\, dF_0(\lambda)\, dF_0(\mu)\, ds_1\, dt_1$$

If we choose $\psi(\lambda,\mu) = \varphi(\lambda,\mu)f_0(\lambda)f_0(\mu)$, then the condition that $\varphi \in L^2(F_0 \otimes F_0)$ is equivalent to

$$\iint_{\mathbb{R}^d \times \mathbb{R}^d} \frac{|\psi(\lambda,\mu)|^2}{(f_0(\lambda)f_0(\mu))^2} f_0(\lambda)f_0(\mu)\, d\lambda\, d\mu < \infty.$$

□

Let us now recall that if $\varphi(\lambda)$ is a Fourier transform of a square-integrable function $a(t)$,

$$\varphi(\lambda) = \int_{\mathbb{R}^d} e^{-i\langle \lambda,t\rangle_{\mathbb{R}^d}} a(t)\, dt$$

and $(a_1 * a_2)(t)$ denotes the *convolution* of two functions $a_1(t), a_2(t)$,

$$(a_1 * a_2)(t) = \int_{\mathbb{R}^d} a_1(s)a_2(t-s)\, ds,$$

then

$$\varphi_1(\lambda)\varphi_2(\lambda) = \int_{\mathbb{R}^d} e^{-i\langle \lambda,t\rangle_{\mathbb{R}^d}} (a_1 * a_2)(t)\, dt,$$

and

$$(a_1 * a_2)(t) = \frac{1}{(2\pi)^d} \int_{\mathbb{R}^d} e^{i\langle \lambda,t\rangle_{\mathbb{R}^d}} \varphi_1(\lambda)\varphi_2(\lambda)\, d\lambda.$$

Theorem 6.4.4. *Assume that there exist constants $0 < c_1 \le c_2 < \infty$ such that*

$$c_1|\varphi(\lambda)|^2 \le f_0(\lambda) \le c_2|\varphi(\lambda)|^2,$$

we will denote this fact by $0 < f_0(\lambda) \asymp |\varphi(\lambda)|^2$. In addition, assume that φ is a Fourier transform of some square integrable function vanishing outside a bounded set $T \subset \mathbb{R}^d$ Let

$$h(\lambda) = \frac{f_1(\lambda) - f_0(\lambda)}{f_0(\lambda)}.$$

If for some constant $M > 0$,

$$\int_{|\lambda| > M} |h(\lambda)|^2 \, d\lambda < \infty$$

then $P_0 \equiv P_1$.

Proof. Let us first assume that $f_0(\lambda) = |\varphi(\lambda)|^2$ and

$$\varphi(t) = \int_{\mathbb{R}^d} e^{i\langle t, \lambda \rangle_{\mathbb{R}^d}} c(\lambda) \, d\lambda,$$

where $c(\lambda) = 0$ on T^c (the complement of T).

Then one can show that there exists a function $a(t) \in L^2(\mathbb{R}^d)$, such that

$$h(\lambda) = \int_{\mathbb{R}^d} e^{i\langle \lambda, t \rangle_{\mathbb{R}^d}} a(t) \, dt,$$

and

$$
\begin{aligned}
b(s,t) &= \int_{\mathbb{R}^d} e^{i\langle \lambda, s-t \rangle_{\mathbb{R}^d}} \left(f_1(\lambda) - f_0(\lambda) \right) d\lambda \\
&= \int_{\mathbb{R}^d} e^{i\langle \lambda, s-t \rangle_{\mathbb{R}^d}} h(\lambda) |\varphi(\lambda)|^2 \, d\lambda,
\end{aligned}
$$

where

$$|\varphi(\lambda)|^2 = \int_{\mathbb{R}^d} e^{i\langle \lambda, t \rangle_{\mathbb{R}^d}} \int_{\mathbb{R}^d} c(u) \overline{c(u-t)} \, du.$$

Let

$$c_0(t - t_1) = \int_{\mathbb{R}^d} e^{i\langle \lambda, t-t_1 \rangle_{\mathbb{R}^d}} \, dF_0(\lambda).$$

Consider

$$c_0(t-t_1) c_0(s-s_1) = \iint_{\mathbb{R}^d \times \mathbb{R}^d} e^{i\langle (t-t_1), \lambda \rangle_{\mathbb{R}^d}} e^{i\langle (s-s_1), \mu \rangle_{\mathbb{R}^d}} \, dF_0(\lambda) dF_0(\mu),$$

and an RKHS $K(C_0 \otimes C_0)$ consisting of functions $\hat{\varphi}$ of the form

$$\hat{\varphi}(s,t) = \iint_{\mathbb{R}^d \times \mathbb{R}^d} e^{i\left(\langle t, \lambda \rangle_{\mathbb{R}^d} \langle s, \mu \rangle_{\mathbb{R}^d} \right)} \varphi(\lambda, \mu) \, dF_0(\lambda) dF_0(\mu),$$

with $\varphi \in L^2(F_0 \otimes F_0)$.

Also

$$\hat{\varphi}(s,t) = \langle \hat{\varphi}(\cdot), C_0 \otimes C_0((s,t), \cdot) \rangle_{K(C_0 \otimes C_0)}$$

As we observed before in Theorem 6.4.3, it follows from Theorem 6.2.2 (c) that, under our assumptions, $P_0 \equiv P_1$ if and only if $b(s,t) = C_1(s,t) - C_0(s,t) \in K(C_0 \otimes C_0)$. Now, for $s,t \in D$,

$$b(s,t) = \iint_{\mathbb{R}^d \times \mathbb{R}^d} e^{-i\left(\langle t,\lambda \rangle_{\mathbb{R}^d} + \langle s,\mu \rangle_{\mathbb{R}^d}\right)} \varphi(\lambda,\mu) f_0(\lambda) f_0(\mu) \, d\lambda \, d\mu.$$

Suppose that $b(s,t)$ can be extended to a square-integrable function on $\mathbb{R}^d \times \mathbb{R}^d$, whose Fourier transform is given by $\psi(\lambda,\mu) = \varphi(\lambda,\mu) f_0(\lambda) f_0(\mu)$, where

$$\iint_{\mathbb{R}^d \times \mathbb{R}^d} \frac{|\psi(\lambda,\mu)|^2}{f_0(\lambda) f_0(\mu)} \, d\lambda \, d\mu < \infty.$$

In our case

$$
\begin{aligned}
b(s,t) &= (2\pi)^d \int_{\mathbb{R}^d} a(v) \int_{\mathbb{R}^d} c(u)\overline{c(-(s-t)-(v-u))} \, du \, dv \\
&= (2\pi)^d \iint_{\mathbb{R}^d \times \mathbb{R}^d} a(u-v) c(s-u)\overline{c(t-v)} \, du \, dv.
\end{aligned}
$$

The functions $c(s-u), c(t-v)$ vanish for u,v outside a compact set T', hence

$$b(s,t) = (2\pi)^d \iint_{T' \times T'} a(u-v) c(s-u)\overline{c(t-v)} \, du \, dv.$$

We now extend the function $a(u-v)$ to a square-integrable function $a(u,v)$ on the entire $\mathbb{R}^d \times \mathbb{R}^d$, so that $a(u,v) = a(u-v)$ on $T' \times T'$. Let $\psi(\lambda,\mu)$ be Fourier transform of $a(u,v)$. Then the function

$$b(s,t) = (2\pi)^d \iint_{\mathbb{R}^d \times \mathbb{R}^d} a(u,v) c(s-u)\overline{c(t-v)} \, du \, dv$$

is an extension of $b(s,t)$ defined earlier to $\mathbb{R}^d \times \mathbb{R}^d$. Its Fourier transform is $\psi(\lambda,\mu)\varphi(\lambda)\varphi(\mu)$, and

$$\iint_{\mathbb{R}^d \times \mathbb{R}^d} \frac{|\psi(\lambda,\mu)\varphi(\lambda)\overline{\varphi(\mu)}|^2}{f_0(\lambda) f_0(\mu)} \, d\lambda \, d\mu = \iint_{\mathbb{R}^d \times \mathbb{R}^d} |\psi(u,v)|^2 \, du \, dv < \infty.$$

$$\iint_{\mathbb{R}^d \times \mathbb{R}^d} \frac{|\psi(\lambda,\mu)\varphi(\lambda)\overline{\varphi(\mu)}|^2}{f_0(\lambda) f_0(\mu)} \, d\lambda \, d\mu = \iint_{\mathbb{R}^d \times \mathbb{R}^d} |\psi(u,v)|^2 \, du \, dv < \infty.$$

Now let us consider $f_0(\lambda) \asymp |\varphi(\lambda)|^2$ and $f_1(\lambda) > f_0(\lambda)$, and $\tilde{f}_1(\lambda) = |\varphi(\lambda)|^2$. Let

$$\tilde{f}_1(\lambda) = \tilde{f}_0(\lambda) + (f_1(\lambda) - f_0(\lambda)).$$

Arguing as above, we can extend the function $b(s,t)$

$$b(s,t) = \int_{\mathbb{R}^d} e^{i\langle \lambda, s-t \rangle_{\mathbb{R}^d}} \left(\tilde{f}_1(\lambda) - \tilde{f}_0(\lambda) \right) d\lambda$$

$$= \int_{\mathbb{R}^d} e^{i\langle \lambda, s-t \rangle_{\mathbb{R}^d}} |f_1(\lambda) - f_0(\lambda)| \, d\lambda,$$

so that its Fourier transform $\varphi(\lambda, \mu)$ satisfies

$$\iint_{\mathbb{R}^d \times \mathbb{R}^d} \frac{|\varphi(\lambda, \mu)|^2}{\tilde{f}_0(\lambda) \tilde{f}_0(\mu)} \, d\lambda \, d\mu < \infty.$$

For M large enough, there exists a constant γ_3, such that $\tilde{f}_0(\lambda) < \gamma_3 f_0(\lambda)$, for $|\lambda| > M$. Then

$$\int_{|\lambda| > M} \int_{|\lambda| > M} \frac{|\varphi(\lambda, \mu)|^2}{f_0(\lambda) f_0(\mu)} \, d\lambda \, d\mu < \infty$$

implies equivalence $P_1 \equiv P_0$. If $f_1(\lambda) > f_0(\lambda)$, we choose $f_2(\lambda) = f_0(\lambda) + \max(0, f_1(\lambda) - f_0(\lambda))$, then $f_2(\lambda) > f_0(\lambda)$ and $f_2(\lambda) \geq f_1(\lambda)$, and for some constant M_2

$$\int_{|\lambda| > M_2} \left(\frac{f_2(\lambda) - f_0(\lambda)}{f_0(\lambda)} \right)^2 d\lambda < \infty$$

and

$$\int_{|\lambda| > M_2} \left(\frac{f_2(\lambda) - f_1(\lambda)}{f_1(\lambda)} \right)^2 d\lambda < \infty.$$

Let P_2 be a probability measure induced by a Gaussian random field with spectral density f_2. Then the previous argument shows that

$$P_0 \equiv P_2 \quad \text{and} \quad P_1 \equiv P_2,$$

giving $P_0 \equiv P_1$ □

6.5 Absolute Continuity of Gaussian Measures under Translations

We now give an extension of a result on absolute continuity of the Gaussian measure under a translation by a particular type of non-linear random function with values in an RKHS. This is a generalization of the Girsanov theorem [74], which in turn was an extension of the original result of Cameron and Martin [10]. Under the non-anticipativity of a non-linear functional as in the Girsanov theorem, the problem can be handled using martingales. For a more general class of non-linear functionals, Ramer [107] and Kusuoka [70] first proved a Girsanov-type theorem generalizing earlier work of Cameron and Martin [10], Gross [40], [41], and Kuo [68] for Gaussian measures on Abstract Wiener Space (AWS). Alternate approach for Brownian motion was proposed by Buckdahn [8], [9], where he starts from a finite-dimensional case using uniform integrability of the density and a Novikov-type condition. All three authors, Ramer, Kusuoka, and Buckdahn, give conditions on the random shift from which finite dimensional assumptions follow. However, Buckdahn uses a clever approximation by "elementary functions" to show the sufficiency of the Novikov-type condition.

Let $X = \{X_t, t \in T\}$ be a Gaussian process with covariance C, $K(C)$ denote the Reproducing Kernel Hilbert Space of C, and μ be the Kolmogorov measure of X on $\mathbb{R}^T$. We assume throughout that μ is Radon with locally convex support $\mathcal{X} \subseteq \mathbb{R}^T$. Since $\mathbb{R}^T$ is a locally convex topological vector space (LCTVS for short), $\mathcal{X}$ is an LCTVS under the induced topology from $\mathbb{R}^T$. By a result of Borel [7], the injection $i : K(C) \to \mathcal{X}$ is continuous and the closure $\overline{i(K(C))} = \mathcal{X}$. Here both $K(C)$ and $\mathcal{X}$ are separable. We need the following result from [31].

Proposition 6.5.1. *Let μ be a Gaussian measure on $(i, K(C), \mathcal{X})$ as above. Let $K \subseteq \mathcal{X}$ be a finite-dimensional linear subspace with $K \subseteq \mathcal{X}^*$ and $\{k_1, k_2, \ldots, k_n\} \subseteq K$, its orthonormal basis. Let $\tilde{K} = \bigcap_{j=1}^{n} \ker(k_j)$ (a closed complement of K in $\mathcal{X}$) and denote by P_K and $P_{\tilde{K}}$, the projections of $\mathcal{X}$ onto K and $\tilde{K}$, respectively. Define $\mu_K = P_K \mu$, $\mu_{\tilde{K}} = P_{\tilde{K}} \mu$, the image measures of μ under P_K and $P_{\tilde{K}}$. Then,*

$$
\int_{\mathcal{X}} f(x)\mu(dx) = \int_{\tilde{K} \times K} f(\tilde{x} + x) \mu_{\tilde{K}}(d\tilde{x}) \otimes \mu_K(dx)
$$

$$
= \int_{\tilde{K} \times R^n} f\left(\tilde{x} + \sum_{j=1}^{n} x_j k_j\right) \mu_{\tilde{K}}(d\tilde{x})
$$

$$
\otimes \frac{1}{2\pi}^{n/2} \exp\left\{-\frac{1}{2}\sum_{j=1}^{n} x_j^2\right\} dx_1 \ldots dx_n,
$$

for any measurable function $f : \mathcal{X} \to \mathbb{R}_+$.

Proof. For any Hausdorff LCTVS $\mathcal{X}$, if F is its finite dimensional subspace, then F and $\tilde{F}$, defined as above, are closed subspaces of $\mathcal{X}$, such that $F \oplus \tilde{F} = \mathcal{X}$. If $x \in \mathcal{X}$, then x can be decomposed in a unique way into $x = x_F + x_{\tilde{F}}$ with $x_F \in F$ and $x_{\tilde{F}} \in \tilde{F}$.

The projections P_F and $P_{\tilde{F}}$ are linear and continuous, since $P_F(x) = x_F = \sum_{j=1}^{n} k_j(x)k_j$ and $P_{\tilde{F}}(x) = x - x_F$. Hence, the image measures $P_F\mu, P_{\tilde{F}}\mu$ are Gaussian measures on F and $\tilde{F}$, respectively.

We want to prove that $\mu = \mu_F \otimes \mu_{\tilde{F}}$ on $F \times \tilde{F} = \mathcal{X}$. First we will prove that $\mu_F \otimes \mu_{\tilde{F}}$ is a Gaussian measure and that the continuous, linear functionals on $\mathcal{X}$ can be decomposed into a sum of two independent (with respect to the measure μ) Gaussian random variables related to the subspaces F and $\tilde{F}$.

To show that $\mu_F \otimes \mu_{\tilde{F}}$ is a Gaussian measure we first prove that for every $\varphi \in \mathcal{X}^*$, $\varphi \circ P_F$ and $\varphi \circ P_{\tilde{F}}$ are independent Gaussian random variables with respect to $\mu_F \otimes \mu_{\tilde{F}}$ on $\mathcal{X}$. To see this, note that

$$
\varphi \circ P_F(x) = \varphi \circ P_F(x_F + x_{\tilde{F}}) = \varphi(x_F).
$$

Hence, $\varphi \circ P_F|_F = \varphi|_F \in F^*$ is normally distributed with respect to μ_F.

Thus, $\varphi \circ P_F$ on $\mathcal{X}$ is also normally distributed with respect to $\mu_F \otimes \mu_{\tilde{F}}$ since the values of this functional are independent of the component belonging to $\tilde{F}$. By the same argument, $\varphi \circ P_{\tilde{F}}$ is Gaussian and $(\varphi \circ P_F, \varphi \circ P_{\tilde{F}})$ is Gaussian. Independence follows from the equalities below:

$$\int_{\mathcal{X}} \varphi \circ P_F(x) \varphi \circ P_{\tilde{F}}(x) \mu_F \otimes \mu_{\tilde{F}}(dx)$$

$$= \int_{F \times \tilde{F}} \varphi \circ P_F(x_F + x_{\tilde{F}}) \varphi \circ P_{\tilde{F}}(x_F + x_{\tilde{F}}) d\mu_F \otimes \mu_{\tilde{F}}$$

$$= \int_{F \times \tilde{F}} \varphi |_F (x_F) \varphi |_{\tilde{F}} (x_{\tilde{F}}) d\mu_F \otimes \mu_{\tilde{F}}$$

$$= \int_F \varphi |_F (x_F) d\mu_F \int_{\tilde{F}} \varphi |_{\tilde{F}} (x_{\tilde{F}}) d\mu_{\tilde{F}}$$

$$= \int_{\mathcal{X}} \varphi \circ P_F(x) d\mu_F \otimes \mu_{\tilde{F}} \int_{\mathcal{X}} \varphi \circ P_{\tilde{F}}(x) d\mu_F \otimes \mu_{\tilde{F}}.$$

Finally, $\varphi = \varphi \circ P_F + \varphi \circ P_{\tilde{F}}$ is a sum of independent Gaussian random variables. Hence, it is itself a Gaussian random variable with respect to $\mu_F \otimes \mu_{\tilde{F}}$ on $\mathcal{X}$.

Next we show that $\varphi \circ P_F$ and $\varphi \circ P_{\tilde{F}}$ are independent Gaussian random variables relative to μ on $\mathcal{X}$.

For every $\varphi \in \mathcal{X}^*$, we have $\varphi(x) = \sum_{i=1}^{\infty} (\varphi, e_i) I_1(e_i)(x)$, for an ONB $\{e_i\}_{i=1}^{\infty}$ in K, where $\{e_i\}_{i=1}^{\infty} \subseteq \mathcal{X}^*$ and $e_i = k_i \in F$, $(i = 1, \ldots n)$ (see [84]). Therefore, we can express compositions of the functional φ with projections P_F and $P_{\tilde{F}}$ as follows:

$$\varphi \circ P_F(x) = \sum_{i=1}^{n} (\varphi, e_i) e_i(x)$$

$$\varphi \circ P_{\tilde{F}}(x) = \sum_{i=1}^{\infty} (\varphi, e_i) e_i(x - \sum_{j=1}^{n} e_i(x) e_i)$$

$$= \sum_{i=1}^{\infty} (\varphi, e_i) e_i(x) - \sum_{i=1}^{n} (\varphi, e_i) e_i(x)$$

$$= \sum_{i=n+1}^{\infty} (\varphi, e_i) e_i(x)$$

(even though x need not be equal to $\sum_{i=1}^{\infty} e_i(x) e_i$).

As $\{e_i(x) = I_1(e_i)(x)\}_{i=1}^{n}$ and $\{e_i(x)\}_{i=n+1}^{\infty}$ are independent families of random variables with respect to μ, $\varphi \circ P_F$ and $\varphi \circ P_{\tilde{F}}$ are independent.

Now to prove that $\mu = \mu_F \otimes \mu_{\tilde{F}}$, we compare the characteristic functionals of these measures.

$$\mu_F \hat{\otimes} \mu_{\tilde{F}}(\varphi)$$

$$- \int_{\mathcal{X}} \exp\{i\psi(x)\} \mu_F \otimes \mu_{\tilde{F}}(dx)$$

$$= \int_{\mathcal{X}} \exp\{i(\varphi \circ P_F(x) + \varphi \circ P_{\tilde{F}}(x))\} \mu_F \otimes \mu_{\tilde{F}}(dx)$$

$$= \int_F \exp\{i\varphi\,|_F\,(x_F)\}\mu_F(dx_F) \int_{\tilde{F}} \exp\{i\varphi\,|_{\tilde{F}}\,(x_{\tilde{F}})\}\mu_{\tilde{F}}(dx_{\tilde{F}})$$

$$= \int_{\mathcal{X}} \exp\{i\varphi \circ P_F(x)\}\mu(dx) \int_{\mathcal{X}} \exp\{i\varphi \circ P_{\tilde{F}}(x)\}\mu(dx)$$

$$= \int_{\mathcal{X}} \exp\{i\varphi(x)\}\mu(dx) = \hat{\mu}(\varphi)$$

As $\mu = \mu_F \otimes \mu_{\tilde{F}}$, we have that

$$\int_{\mathcal{X}} f(x)\mu(dx) = \int_{F \times \tilde{F}} f((x_F + x_{\tilde{F}}))\mu_F \otimes \mu_{\tilde{F}}(dx_F, dx_{\tilde{F}})$$

for any measurable function $f : \mathcal{X} \to \mathbb{R}_+$. $\qquad\qquad\qquad\qquad\qquad\square$

The functions we shall consider for a shift will be stochastically integrable in the sense of Ramer (see [107]) and we recall some concepts related to the *Ramer stochastic integral*

Definition 6.5.1. *Let μ be a Gaussian measure on $(i, H = K(C), \mathcal{X})$ as above. A Bochner measurable map G from $\mathcal{X}$ to $H \subseteq \mathcal{X}$ is said to be stochastic Gateaux H-differentiable (SGD) if there exists a Bochner measurable map $\tilde{D}G : \mathcal{X} \to L(H, H)$, so that for each $h, h' \in H$,*

$$\frac{1}{t} \left\langle h', G(x + th) - G(x) \right\rangle_H \to \left\langle h', \tilde{D}G(x)h \right\rangle_H \quad \text{in probability } \mu \text{ as } t \to 0.$$

We call $\tilde{D}G$ the stochastic Gateaux H-derivative of G.

Definition 6.5.2. (a) *A function $f : \mathbb{R} \to H$ is called absolutely continuous if for any $-\infty < a < b < \infty$ and $\varepsilon > 0$, there exists some $\delta(\varepsilon, a, b) > 0$ such that $\sum_{i=1}^n \|f(t_i) - f(s_i)\|_H < \varepsilon$ holds for any integer n and $a \le t_1 < s_1 \le t_2 < s_2 ... t_n < s_n \le b$, $\sum_{i=1}^n |t_i - s_i| < \delta(\varepsilon, a, b)$.*

(b) *A Bochner measurable function $G : \mathcal{X} \to H$ is called ray absolutely continuous (RAC), if for every $h \in H$, there exists a Bochner measurable map $G_h : \mathcal{X} \to H$, so that $\mu\{G_h = G\} = 1$ and $f(t) = G_h(x + th)$ is absolutely continuous in t for each $x \in \mathcal{X}$.*

We denote the class of functions $G : \mathcal{X} \to H$ which are SGD and RAC by $H^1(\mathcal{X} \to H, d\mu)$. Let $\mathcal{P}(H)$ be the set of all finite dimensional projections on H and $\mathcal{P}^*(H)$ be the subset of $\mathcal{P}(H)$ consisting of projections with range in $\mathcal{X}^*$. We now define the Itô–Ramer integral [31], [70], [107].

Definition 6.5.3. *A (Bochner) measurable function $G : \mathcal{X} \to H$ is called Itô–Ramer integrable if*

(a) $G \in H^1(\mathcal{X} \to H, d\mu)$.

(b) $(\tilde{D}G)(x) \in H^{\otimes 2}$ μ-a.e.

(c) *There exists a measurable function $\hat{L}G$ such that*

$$L_P G(x) = \langle PG(x), x \rangle - tr\tilde{P}\tilde{D}G(x) \overset{\mu}{\to} \hat{L}G(x)$$

as $P \to I_H$ (identity operator on H), $P \in \mathcal{P}^*(H)$.

We call LG the Itô–Ramer integral of G and denote by $\mathcal{D}(L)$ the class of Itô-Ramer integrable functions.

We will now study equivalence under non-linear transformation. For A in $H^{\otimes 2}$, we introduce the *Carleman–Fredholm determinant* of A,

$$d_c(I_H - A) = \prod_{i=1}^{\infty} (1 - \lambda_i) \exp(\lambda_i)$$

where $\{\lambda_i\}_{i=1}^{\infty}$ are eigenvalues of A.

The reader will find a wealth of information on determinants in [38], we state the properties useful in our exposition as an exercise.

Exercise 6.5.1. *Prove the following properties of the Carleman–Fredholm determinant for $A, B \in \mathcal{L}_2(H)$,*

(a) $|d_c(I_H - A)| \leq \exp\left(\frac{1}{2}\|A\|^2_{\mathcal{L}_2(H)}\right)$,

(b)

$$|d_c(I_H - A) - d_c(I_H - B)| \leq \|A - B\|_{\mathcal{L}_2(H)} \exp\left(\frac{1}{2}\left(\|A\|_{\mathcal{L}_2(H)} + \|B\|_{\mathcal{L}_2(H)} + 1\right)^2\right),$$

(c) *the Carleman–Fredholm determinant is a continuous function on $\mathcal{L}_2(H)$,*

(d) $d_c(I_H - A) = \det(I_H - A)\operatorname{tr}(A)$ *for $A \in \mathcal{L}_1(H)$,*

(e) $d_c((I_H - A)(I_H - B)) = d_c(I_H - A)d_c(I_H - B)\exp(-\operatorname{tr}(AB))$.

Let $F : \mathcal{X} \to H$ be a Bochner measurable transformation, and denote $F_n = P_n F$, where $P_n \in \mathcal{P}(H)$ is a projection onto an n-dimensional subspace of H for each n. For $F \in \mathcal{D}(L)$, let

$$d(x, F) = d_c\left(I_H - \tilde{D}F(x)\right)\exp\left(LF(x) - \frac{1}{2}\|F(x)\|^2_H\right).$$

Remark 6.5.1. *We observe that if $P_n \to I_H$, $(P_n \in \mathcal{P}^*(H))$ then by continuity of the Carleman–Fredholm determinant in $H^{\otimes 2}$ (regarded as the space of Hilbert-Schmidt operators on H) and by the definition of the Itô-Ramer integral of F (Definition 6.5.3), we have the following convergence:*

$$d(x, F_n) \to d(x, F)$$

in probability measure μ.

In order to proceed to the central result of this section we need to introduce the transformation theorem on $\mathbb{R}^n$ for Lipschitz continuous transformations. The theorem of Rademacher [24], [25] states that a Lipschitz transformation is a.e. differentiable.

Theorem 6.5.1 (Rademacher). *Let $f : \mathbb{R}^n \to \mathbb{R}^m$ be a Lipschitz transformation. Then f is λ–a.e. differentiable.*

The general form of the transformation theorem considers $f : \mathbb{R}^n \to \mathbb{R}^m$, $n \le m$, and uses the n-dimensional Hausdorff measure on $\mathbb{R}^m$. When $n = m$ the Hausdorff and Lebesgue measures coincide ([25], 2.10.35).

Theorem 6.5.2. *The Lebesgue and Hausdorff measures on $\mathbb{R}^n$ coincide.*

Now, let $f = (f_1,...,f_n) : \mathbb{R}^n \to \mathbb{R}^m$ be Lipschitz. By Rademacher's Theorem 6.5.1, f is λ–a.e. differentiable, so that, for λ–a.e. $x \in \mathbb{R}^n$, there exists $Df(x) \in \mathcal{L}(\mathbb{R}^n, \mathbb{R}^m)$,

$$Df(x) = \left(\frac{\partial f_i(x)}{\partial x_j}\right)_{i,j}, \quad i = 1,...,m; \; j = 1,...,n.$$

The Jacobian of the transformation f is defined as $Jf(x) = |\det Df(x)|$. The following change of variable formula, Theorem 2, Section 3.3.3, is presented in [24].

Theorem 6.5.3. *Let $f =: \mathbb{R}^n \to \mathbb{R}^m$, $n \le m$, be Lipschitz. Then for each function $g \in L^1(\mathbb{R}^n)$,*

$$\int_{\mathbb{R}^n} g(x)Jf(x)\,dx = \int_{\mathbb{R}^n}\left(\sum_{x\in f^{-1}(y)} g(x)\right)dy. \tag{6.15}$$

By defining $g(x) = h(f(x))$ we obtain the following form of the transformation theorem which will be useful for us.

Theorem 6.5.4. *Let $f : \mathbb{R}^n \to \mathbb{R}^n$ be bijective Lipschitz transformation and $h : \mathbb{R}^n \to \mathbb{R} \in L^1(\mathbb{R}^n)$. Then*

$$\int_{\mathbb{R}^n} h(f(x))Jf(x)\,dx = \int_{f(\mathbb{R}^n)} h(y)\,dy. \tag{6.16}$$

We present the main theorem.

Theorem 6.5.5. *Let $F : \mathcal{X} \to H$ be a Bochner measurable function with the following properties:*

(a) $F \in \mathcal{D}(L)$,

(b) *For each n, $(I_{\mathcal{X}} - F_n)$ is bijective on $\mathcal{X}$ and for all $\tilde{x} \in \widetilde{P_n\mathcal{X}}$, the closed complement of $P_n\mathcal{X}$ in $\mathcal{X}$, the mapping $z \to z - F_n(\tilde{x}+z)$ from $P_n\mathcal{X}$ to $P_n\mathcal{X}$ satisfies the change of variable formula (6.16) on $P_n\mathcal{X}$ for some $\{P_n\} \subseteq \mathcal{P}^*(H)$ with $P_n \to I_{\mathcal{H}}$.*

(c) $\{d(x,F_n)\}_{n=1}^{\infty}$ *is uniformly integrable.*

Then for all bounded measurable functions g on $\mathcal{X}$,

$$\int_{\mathcal{X}} g((I_{\mathcal{X}} - F)(x))|d(x,F)|\,\mu(dx) = \int_{\mathcal{X}} g(x)\mu(dx). \tag{6.17}$$

Proof. With $K = P_n \mathcal{X}$ in Proposition 6.5.1,

$$\int_{\mathcal{X}} g((I_{\mathcal{X}} - F_n)(x)) |d(x, F_n)| \mu(dx)$$

$$= \int_{\tilde{K} \times K} g((I_{\mathcal{X}} - F_n)(\tilde{z} + z)) |d(\tilde{z} + z, F_n)| \mu_{\tilde{K}}(d\tilde{z}) \mu_K(dz)$$

$$= \int_{\tilde{K} \times R^n} g\left(\tilde{z} + (I_K - F_n(\tilde{z} + \cdot))\left(\sum_{j=1}^n x_j e_j\right)\right)$$

$$\times \left|d\left(\tilde{z} + \sum_{j=1}^n x_j e_j, P_n F\right)\right| \mu_{\tilde{K}}(d\tilde{z}) \left(\frac{1}{2\pi}\right)^{\frac{n}{2}} e^{-\frac{1}{2}\sum_{j=1}^n x_j^2} dx_1 \ldots dx_n$$

$$\int_{\tilde{K} \times R^n} g\left(\tilde{z} + \sum_{j=1}^n (x_j - \psi_j(\tilde{z}, x_1 \ldots x_n)) e_j\right)$$

$$\times \left|d\left(\tilde{z} + \sum_{j=1}^n x_j e_j, P_n F\right)\right| \mu_{\tilde{K}}(d\tilde{z}) \left(\frac{1}{2\pi}\right)^{\frac{n}{2}} e^{-\frac{1}{2}\sum_{j=1}^n x_j^2} dx_1 \ldots dx_n,$$

where

$$\psi_i(\tilde{z}, \underline{x}) = \left\langle F\left(\tilde{z} + \sum_{j=1}^n x_j e_j\right), e_i\right\rangle_H, \quad \underline{x} \in R^n.$$

Note that

$$\left|d\left(\tilde{z} + \sum_{j=1}^n x_j e_j, P_n F\right)\right| e^{-\frac{1}{2}\sum_{j=1}^n x_j^2}$$

$$= \left|d_c\left(I_H - DP_n F\left(\tilde{z} + \sum_{j=1}^n x_j e_j\right)\right)\right| \exp\left((LP_n F)\left(\tilde{z} + \sum_{j=1}^n x_j e_j\right)\right.$$

$$\left. -\frac{1}{2}\left(\left\|P_n F\left(\tilde{z} + \sum_{j=1}^n x_j e_j\right)\right\|_H^2 + \sum_{j=1}^n x_j^2\right)\right)$$

$$= |\det(I_{R^n} - D\Psi(\tilde{z}, \underline{x}))| \exp\left(-\frac{1}{2}\sum_{j=1}^n (x_j - \psi_j(\tilde{z}, \underline{x}))^2\right)$$

where

$$D\Psi(\tilde{z}, \underline{x}) = \left(\frac{\partial \psi_i(\tilde{z}, \underline{x})}{\partial x_j}\right)_{i,j=1}^n.$$

Hence

$$\int_{\mathcal{X}} g((I_{\mathcal{X}} - P_n F))(x) |d(x, P_n F)| \mu(dx)$$

$$= \int_{\tilde{K} \times R^n} g\left(\tilde{z} + \sum_{j=1}^n (x_j - \psi_j(\tilde{z}, x)) e_j\right) \left(\frac{1}{2\pi}\right)^{\frac{n}{2}} \exp\left(-\frac{1}{2}\sum_{j=1}^n (x_j - \psi_j(\tilde{z}, \underline{x}))^2\right)$$

$$\times \left| \det \left(I_{R^n} - D\Psi \left(\tilde{z}, \underline{x} \right) \right) \right| dx_1 \ldots dx_n \, \mu_{\tilde{K}} \left(d\tilde{z} \right).$$

By Proposition 6.5.1, using the change of variable formula (6.16) on $\mathbb{R}^n$, the above equals

$$\int_{\tilde{K} \times R^n} g \left(\tilde{z} + \sum_{j=1}^{n} y_j e_j \right) \left(\frac{1}{2\pi} \right)^{\frac{n}{2}} \exp \left(-\frac{1}{2} \sum_{j=1}^{n} y_j^2 \right) dy_1 \ldots dy_n \, \mu_{\tilde{K}} \left(d\tilde{z} \right)$$

$$= \int_{\mathcal{X}} g(x) \mu(dx)$$

giving

$$\int_{\mathcal{X}} g \left((I_{\mathcal{X}} - F_n)(x) \right) \left| d(x, F_n) \right| \mu(dx) = \int_{\mathcal{X}} g(x) \mu(dx). \tag{6.18}$$

We also observe that it suffices to prove (6.17) for a bounded continuous function g. As $(I_{\mathcal{X}} - P_n F)(x) \to (I_{\mathcal{X}} - F)(x)$, we get $g \left((I_{\mathcal{X}} - P_n F)(x) \right) \to g \left((I_{\mathcal{X}} - F)(x) \right)$ since g is continuous, and because $d(x, F_n)$ is uniformly integrable, $g \left((I_{\mathcal{X}} - F_n)(x) \right) d(x, F_n)$ is uniformly integrable and converges to

$$g(I_{\mathcal{X}} - F)(x) d(x, F) \quad \text{in probability } \mu$$

by Remark 6.5.1. Hence, by (6.18), we obtain the result for a continuous and bounded function g. $\qquad\square$

Exercise 6.5.2. *Assume (a) and (b) in Theorem 6.5.5 and show that*

$$\int_{\mathcal{X}} g \left((I_{\mathcal{X}} - F)(x) \right) \left| d(x, F) \right| \mu(dx) \leq \int_{\mathcal{X}} g(x) \mu(dx).$$

Hint: Use the Fatou lemma.

Corollary 6.5.1. *If $I_{\mathcal{X}} - F$ is bijective and assumptions (a), (b), (c) of Theorem 6.5.5 are satisfied, then*

$$\int_{\mathcal{X}} g(x) \left| d(x, F) \right| \mu(dx) = \int_{\mathcal{X}} g \left((I_{\mathcal{X}} - F)^{-1} x \right) \mu(dx). \tag{6.19}$$

Proof. This follows from (6.17) by observing that $g \circ (I_{\mathcal{X}} - F)^{-1}$ is a bounded measurable function as $\mathcal{X}$ is a Polish space. $\qquad\square$

Corollary 6.5.2. *Let $T = (I_{\mathcal{X}} - F)$ be bijective and assumptions (a), (b), (c) of Theorem 6.5.5 be satisfied, then*

(a) $\mu \circ T \ll \mu$ *and*

$$\frac{d\mu \circ T}{d\mu} = \left| d(x, F) \right|. \tag{6.20}$$

(b) *If* $d_c\left(I_H - \tilde{D}F\right) \neq 0$ μ–*a.e., then* $\mu \circ T \sim \mu$ *and* $\mu \circ T^{-1} \sim \mu$, *and we have the following formula ([97] and [54]):*

$$\frac{d\mu \circ T}{d\mu}(x) = \left(\frac{d\mu \circ T^{-1}}{d\mu}(T(x))\right)^{-1} \tag{6.21}$$

Proof. Part (a) follows from Corollary 6.5.1 and (b) is true since T is bijective.

$\square$

We now give sufficient conditions for uniform integrability and for assumption (b) of Theorem 6.5.5 to hold true, in terms of conditions on the function F. For condition (b) we provide an extension of some analytic work of Kusuoka [70]. Once the result of the next lemma is established, Theorem 6.5.4 will allow us to use the change of variable formula (6.16).

Lemma 6.5.1. *Let* $F : \mathcal{X} \to H$ *be a Bochner measurable transformation with* $F \in \mathcal{D}(L)$ *and assume that there exists a constant* $0 < c < 1$, *such that*

$$\|F(x+h) - F(x)\|_H \leq c\|h\|_H \tag{6.22}$$

for all $x \in \mathcal{X}$, $h \in H$ *(such a transformation* F *is called an* H-*contraction). Then* $(I_{\mathcal{X}} - F) : \mathcal{X} \to \mathcal{X}$ *is bijective and* $d_c(I_H - \tilde{D}F) \neq 0$ μ–*a.e.*

In particular, for any $P \in \mathcal{P}(H)$, $(I_{\mathcal{X}} - PF)$ *is bijective,* $d_c\left(I_H - \tilde{D}PF\right) \neq 0$ μ–*a.e. and for all* $x \in \mathcal{X}$, *the mapping* $z \to z - PF(x+z)$ *on* $P\mathcal{X}$ *is Lipschitz continuous and homeomorphic.*

Proof. This proof is a simple modification of the proof of Theorem 6.1 in [70] and is provided for the reader's convenience. For any $x \in \mathcal{X}$ we define inductively

$$\begin{cases} u_0(x) = 0, \\ u_{n+1}(x) = F\left(x + u_n(x)\right), & n = 1, 2, \ldots \end{cases}$$

Since F is contractive we have

$$\begin{aligned} \|u_{n+1}(x) - u_n(x)\|_H &\leq c\|u_n(x) - u_{n-1}(x)\|_H \\ &\leq c^n\|F(x)\|_H. \end{aligned}$$

It follows that $u_n(x)$ is a Cauchy sequence, hence there exists $u(x) = \lim_{n\to\infty} u_n(x)$. Since $F(x+u(x)) = u(x)$, we have

$$(I_{\mathcal{X}} - F)(x + u(x)) = x,$$

showing that $I_{\mathcal{X}} - F : \mathcal{X} \to H$ is surjective.

To show that F is also injective, suppose that $(I_{\mathcal{X}} - G)x_1 = (I_{\mathcal{X}} - G)x_2$ for some $x_1, x_2 \in \mathcal{X}$. Then $x_1 - x_2 = G(x_1) - G(x_2) \in H$, and

$$\|x_1 - x_2\|_H = \|F(x_2 + x_1 - x_2) - F(x_2)\|_H \leq c\|x_1 - x_2\|_H,$$

proving that $x_1 = x_2$.

To prove that $d_c(I_H - \tilde{D}F) \neq 0$ μ–a.e. we note that the eigenvalues $\lambda_k(x)$ of $\tilde{D}F(x)$ are square summable and $|\lambda_k(x)| \leq c < 1$. It is left to the reader in Exercise 6.5.3 to show that $\prod_{k=1}^{\infty} (1 - \lambda_k) e^{\lambda_k} > 0$.

Since the function $PF : \mathcal{X} \to H$ satisfies all conditions of the lemma imposed on F, the last claim of the lemma follows. $\square$

Exercise 6.5.3. *Show that if $A \in \mathcal{L}_2(H)$ with $\|A\|_{L(H)} < 1$, then $d_c(I_H - A) > 0$.*

We recall a stronger concept of differentiation, the Fréchet H-derivative.

Definition 6.5.4. *A map $F : \mathcal{X} \to H$ is H-Fréchet differentiable at $x \in \mathcal{X}$ if there exists a linear operator $D^{\mathcal{F}} F(x) : H \to H$, such that*

$$\left\| F(x+h) - F(x) - D^{\mathcal{F}} F(x)h \right\|_H = o(\|h\|_H) \quad as \ \|h\|_H \to 0.$$

It will be convenient to use the following definition.

Definition 6.5.5. *We say that a measurable map $F : \mathcal{X} \to H$ is in class $H - C^1$ if for all $x \in \mathcal{X}$ F is H-Fréchet differentiable with the H-Fréchet derivative $D^{\mathcal{F}} F(x) \in H^{\otimes 2}$, and the map $D^{\mathcal{F}} F(x + \cdot) : H \to H^{\otimes 2}$ is continuous.*

The importance of $H - C^1$ maps is highlighted by the fact that $H - C^1$ functions are Itô–Ramer integrable. We will now prove this assertion. We need the following result, Lemma 4.1 from [107].

Lemma 6.5.2. *Let $f : \mathbb{R}^n \to \mathbb{R}^n$, be a stochastic $\mathbb{R}^n$-Gateaux differentiable map (w.r.t. γ_n). Assume that $\|f\|_{\mathbb{R}^n}$ and $\left\| \tilde{D}f \right\|_2 \in L^2(\mathbb{R}^n, \gamma_n)$, where γ_n denotes the standard normal distribution, $N(0, I_{\mathbb{R}^n})$, on $\mathbb{R}^n$. Then*

$$\int_{\mathbb{R}^n} \left(\langle f(x), x \rangle_{\mathbb{R}^n} - \mathrm{tr}\tilde{D}f(x) \right)^2 \gamma_n(dx) \tag{6.23}$$

$$\leq \int_{\mathbb{R}^n} \left(\|f(x)\|_{\mathbb{R}^n}^2 + \left\| \tilde{D}f(x) \right\|_{L^2(\mathbb{R}^n, \gamma_n)}^2 \right) \gamma_n(dx).$$

In order to prove our assertion about Itô-Ramer integrability of $H - C^1$ functions, we introduce some analytic work which is an extension of the results of Kusuoka, contained in paragraph 4 of [70]. Kusuoka considered an Abstract Wiener Space (ι, K, B), while we are interested in a more general situation of the triple $(\iota, H, \mathcal{X})$, associated with a Gaussian process.

Definition 6.5.6. *Let $A \subseteq \mathcal{X}$ be any subset. Define a function $\rho(\cdot; A) : \mathcal{X} \to [0, +\infty]$ by :*

$$\rho(x; A) = \begin{cases} \inf\{\|h\|_H; \ x + h \in A\} & if \ (A - x) \cap H \neq \varnothing \\ +\infty & otherwise \end{cases}$$

The following proposition can be proved as in [70] and it is left for the reader as an exercise.

Proposition 6.5.2. **(a)** *If subsets A and A' of $\mathcal{X}$ satisfy $A \subseteq A'$, then for every $x \in \mathcal{X}$,*

$$\rho(x;A) \geq \rho(x;A')$$

(b) *For every $A \subset \mathcal{X}$, $h \in H$, and $x \in \mathcal{X}$*

$$\rho(x+h;A) \leq \|h\|_H + \rho(x;A).$$

(c) *Let $\{A_n\}_{n=1}^{\infty}$ be an increasing sequence of subsets of $\mathcal{X}$ and $A = \bigcup_{n=1}^{\infty} A_n$, then for every $x \in \mathcal{X}$,*

$$\rho(x;A_n) \downarrow \rho(x;A) \quad as\ n \to \infty.$$

Exercise 6.5.4. *Prove Proposition 6.5.2.*

Theorem 6.5.6. **(a)** *If C is a compact subset of $\mathcal{X}$, then $\rho(\cdot;C) : \mathcal{X} \to [0,+\infty]$ is lower semi-continuous.*

(b) *If G is a σ-compact subset of $\mathcal{X}$, then $\rho(\cdot;G) : \mathcal{X} \to [0,+\infty]$ is measurable.*

Proof. Since (b) is a consequence of (a) and Proposition 6.5.2 (c), it is enough to prove (a). We follow the idea of the proof given in [70].

Define $A_a = \{x \in \mathcal{X} : \rho(x,C) \leq a\}$, and let $B(a)$ denote the closed ball in H of radius a centered at 0. We want to show that $A_a = C + B(a)$. The inclusion $A_a \supseteq C + B(a)$ is clear. For the opposite inclusion, we take $x \in A_a$. Then there exists $\{h_n\}_{n=1}^{\infty} \subseteq (C-x) \cap H$, such that $\|h_n\| \leq a + \frac{1}{n}$. Being norm bounded, the sequence $\{h_n\}_{n=1}^{\infty}$ contains a weakly convergent subsequence $\{h_{n_k}\}_{k=1}^{\infty}$. Let $h \in H$ denote its limit. Since $\mathcal{X}^* \subseteq H$ and for all $t \in T$, $x_t(h) = h(t)$ (point evaluation) is an element of $\mathcal{X}^*$, we also have $h_{n_k} \to h$ in $\mathcal{X}$ (the convergence in $\mathcal{X}$ is a pointwise convergence). Also,

$$
\begin{aligned}
\|h\|_H &= \sup\{\langle h,x\rangle_H ; x \in \mathcal{X}^*, \|x\|_H \leq 1\} \\
&= \sup\left\{\lim_{k \to \infty} \langle h_{n_k}, x\rangle_H ; x \in \mathcal{X}^*, \|x\|_H \leq 1\right\} \\
&\leq \limsup_{n \to \infty} \|h_n\|_H \leq a.
\end{aligned}
$$

Thus, $h \in B(a)$. Since $C \subset \mathcal{X}$ is compact and $h_{n_k} \to h$ in $\mathcal{X}$ with $x + h_{n_k} \in C$, $(k = 1, 2, \ldots)$, $x + h \in C$ and therefore, $x \in C + B(a)$. Thus, $A_a = C + B(a)$. Using Lemma 6.5.3 we can see that $B(a) \subseteq X$ is closed; therefore, $A_a = C + B(a) \subseteq \mathcal{X}$ is closed. $\qquad\square$

Lemma 6.5.3. *Let X be a reflexive Banach space and Y be an LCTVS. Let $T : X \to Y$ be linear and continuous. Then $T(B_X(0,1)) \subseteq Y$ is closed, where $B_X(0,1)$ is a closed unit ball centered at 0 in X.*

Proof. The mapping $T : X \to Y$ is linear and continuous, hence $T : X_\omega \to Y_\omega$ is linear and continuous (the subscript ω indicates that a given space is considered with its weak topology). This is because if $\{x_\alpha\}$ is a net in X with $x_\alpha \to x$ in X_ω, then $\forall y^* \in Y^*$, $y^*(Tx_\alpha) = (y^*T)x_\alpha \to (y^*T)x = y^*(Tx)$, as $(y^*T) \in X^*$.

Because $X^{**} \cong X$ by the canonical isomorphism κ, we get that $T \circ \kappa^{-1} :$ $X_\omega^{**} \to Y_\omega$ is linear and continuous and further, $T \circ \kappa^{-1} : X_{\omega-*}^{**} \to Y_\omega$ is linear and continuous (where $\omega - *$ denotes the weak–$*$ topology). The latter holds because the reflexivity of X implies the reflexivity of X^*. Now, the closed unit ball $B_{X^{**}}(0,1)$ is $\omega - *$ compact by Alaoglu-Banach theorem. That means that $\kappa(B_X(0,1))$ is $\omega - *$ compact in X^{**}. Hence, $T \circ \kappa^{-1}(\kappa(B_X(0,1))) =$ $T(B_X(0,1))$ is ω-closed in Y. Because Y and Y_ω have the same closed, convex sets, $T(B_X(0,1))$ is closed in the topology of Y and the lemma is proved. $\square$

The next theorem can be proved as in [70], with obvious changes. It provides a simple condition for a map to be SGD.

Theorem 6.5.7. *Let E be a separable, reflexive Banach space and $F : \mathcal{X} \to E$ be a measurable map and suppose that there exists a constant $c > 0$ such that for every $x \in \mathcal{X}$, and $h \in H$, $\|F(x+h) - F(x)\|_E \le c\|h\|_H$. Then there exists a measurable subset D_0 of $\mathcal{X}$ and a map $DF : \mathcal{X} \to L(H,E)$, such that*

(a) $\mu(D_0) = 1$

(b) $\lim_{t \to 0} \frac{1}{t}(F(x+th) - F(x)) = DF(x)h$, *for every $x \in D_0$, and $h \in H$*

(c) $DF(\cdot)h : \mathcal{X} \to E$ *is measurable for every $h \in H$*

In particular, if $DF : \mathcal{X} \to L(H,E)$ is strongly measurable, then $F \in H^1(\mathcal{X} \to E; d\mu)$.

Corollary 6.5.3. *Let G be a σ-compact subset of $\mathcal{X}$ and ϕ be a smooth function with compact support in $\mathbb{R}$. Then $g(\cdot) = \phi(\rho(\cdot;G)) : \mathcal{X} \to \mathbb{R}$, with the convention that $\phi(\infty) = 0$, belongs to $H^1(\mathcal{X} \to \mathbb{R}; d\mu)$ and*

$$\|Dg(x)\|_H \le \sup\left\{\left|\frac{d\phi}{dt}\right| ; t \in \mathbb{R}\right\} \tag{6.24}$$

for μ–a.e. x.

Proof. First we observe that by Theorem 6.5.6, (b) g is measurable. Also,

$$\|g(x+h) - g(x)\| \le \sup\left\{\left|\frac{d\phi}{dt}(t)\right| ; t \in \mathbb{R}\right\}(\rho(x+h;G) - \rho(x,G))$$
$$\le \sup\left\{\left|\frac{d\phi}{dt}(t)\right| ; t \in R\right\}\|h\|_H$$

by Proposition 6.5.2 (b) (the convention is here that $\infty - \infty = 0$, note that $\rho(x+h,G) = \infty$ if and only if $\rho(x,G) = \infty$).

Therefore, the assumptions of Theorem 6.5.7 are satisfied. $Dg(\cdot)$ can be thought of as a map from $\mathcal{X}$ to H^*. Thus, $Dg : \mathcal{X} \to H^*$ is weakly measurable and therefore it is strongly measurable in view of the separability of H. Inequality (6.24) is obvious. $\square$

Theorem 6.5.8. *Let $g \in H^1(\mathcal{X} \to H; d\mu)$ with $\tilde{D}g \in H^{\otimes 2}$ μ–a.e. Assume that $\|g\|_H$, $\|\tilde{D}g\|_{H^{\otimes 2}} \in L_2(\mathcal{X}, \mu)$. Then $g \in \mathcal{D}(L)$ and*

$$\int |Lg(x)|^2 \, \mu(dx) \leq \int_{\mathcal{X}} \|g(x)\|_H^2 \, \mu(dx) + \int_{\mathcal{X}} \|\tilde{D}g(x)\|_{H^{\otimes 2}}^2 \, \mu(dx).$$

Proof. Any $P_n \in \mathcal{P}^*(H)$ with $\dim P_n(H) = n$ can be written as follows:

$$P_n = \sum_{i=1}^{n} e_i \otimes e_i, \; e_i \in \mathcal{X}^*, \; \{e_i\}_{i=1}^{n} \text{ ONB in } H.$$

First, we will show that $\{L_{P_n}g\}_{n=1}^{\infty}$ is a Cauchy sequence in $L_2(\mathcal{X})$.

$$\int_{\mathcal{X}} \left(L_{P_l}g(x) - L_{P_m}g(x) \right)^2 \mu(dx)$$

$$= \int_{\mathcal{X}} \left\{ \langle (P_l - P_m)g(x), x \rangle - \operatorname{tr}(P_l - P_m)\tilde{D}g(x) \right\}^2 \mu(dx).$$

We can apply Proposition 6.5.1, to obtain that the last expression is equal to

$$\int_{\tilde{F}} \int_{\mathbb{R}^m} \left\{ \left\langle (P_l - P_m)g\left(\sum_{i=1}^{l} \alpha_i e_i + x_{\tilde{F}}\right), \sum_{i=1}^{l} \alpha_i e_i + x_{\tilde{F}} \right\rangle \right.$$

$$\left. + \operatorname{tr}(P_l - P_m)\tilde{D}g\left(\sum_{i=1}^{l} \alpha_i e_i + x_{\tilde{F}}\right) \right\}^2 \mu(dx_{\tilde{F}})d\gamma_l,$$

where we assume that $l \geq m$, $F = \overline{\operatorname{span}}\{e_1, \ldots e_l\}$, $\tilde{F}$, defined as in Proposition 6.5.1, is a closed complement of F in $\mathcal{X}$, $x_{\tilde{F}} = P_{\tilde{F}}x$ and γ_l denotes the standard Gauss measure on $\mathbb{R}^l$. Using Lemma 6.5.2 we can conclude that an upper bound for the last expression is given by

$$\int_{\mathcal{X}} \left\{ \|(P_l - P_m)g(x)\|_H^2 + \|(P_l - P_m)\tilde{D}g(x)\|_{H^{\otimes 2}}^2 \right\} \mu(dx).$$

Both components converge to zero as $l, m \to \infty$. Indeed,

$$\int_{\mathcal{X}} \|(P_l - P_m)g(x)\|_H^2 \mu(dx) = \int_{\mathcal{X}} \left\| \sum_{i=m+1}^{l} (e_i, g(x))_H e_i \right\|_H^2 \mu(dx)$$

$$\leq \int_{\mathcal{X}} \sum_{i=m+1}^{\infty} (e_i, g(x))_H^2 \mu(dx) \to 0,$$

since $g \in L_2(\mathcal{X}, H, d\mu)$.

Similar argument shows that the second component converges to zero. Thus, we have

$$P_n g \to g \in L_2(\mathcal{X}, H, d\mu) \quad \text{and}$$

$$P_n \tilde{D}g \to \tilde{D}g \in L_2(\mathcal{X}, H^{\otimes 2}, d\mu).$$

In addition, one gets the following estimate:

$$\int_{\mathcal{X}} (L_P g(x))^2 \mu(dx) \le \int_{\mathcal{X}} \{ \|Pg(x)\|_H^2 + \|P\tilde{D}g(x)\|_{H^{\otimes 2}}^2 \} \mu(dx) \qquad (6.25)$$

for $P \in \mathcal{P}^*(H)$.

Furthermore, the limit LF, in $L_2(\mathcal{X})$ does not depend on the choice of the sequence of projections. Indeed, let $\{P_n\}_{n=1}^{\infty}$, $\{Q_n\}_{n=1}^{\infty}$ be two sequences in $\mathcal{P}^*(H)$ converging to I_H, then, we have

$$\|L_{P_n}g - L_{Q_m}g\|_{L^2(\mathcal{X})}^2$$

$$\le \|(P_n - Q_m)g(x)\|_{L^2(\mathcal{X},H)}^2 + \|(P_n - Q_m)\tilde{D}g(x)\|_{L^2(\mathcal{X},H^{\otimes 2})}^2$$

$$\le 2\{ \|P_n g - g\|_{L^2(\mathcal{X},H)}^2 + \|P_n \tilde{D}g - \tilde{D}g\|_{L^2(\mathcal{X},H^{\otimes 2})}^2 \}$$

$$+ 2\{ \|Q_m g - g\|_{L^2(\mathcal{X},H)}^2 + \|Q_m \tilde{D}g - \tilde{D}g\|_{L^2(\mathcal{X},H^{\otimes 2})}^2 \}$$

with the RHS converging to zero as $m, n \to \infty$.

The inequality

$$\int_{\mathcal{X}} |Lg(x)|^2 \, \mu(dx) \le \int_{\mathcal{X}} (\|g(x)\|_H^2 + \|\tilde{D}g(x)\|_{H^{\otimes 2}}^2) \mu(dx)$$

follows from (6.25).   □

Theorem 6.5.9. *Let $g \in H^1(\mathcal{X} \to H; d\mu)$ and ω be a positive weight function; that is, $\omega : \mathcal{X} \to \mathbb{R}$ is measurable, $\omega(x) > 0$ for all $x \in \mathcal{X}$ and $\omega(x + \cdot) : H \to \mathbb{R}$ is continuous for all $x \in \mathcal{X}$. Assume that $\tilde{D}g(x) \in H^{\otimes 2}$ for μ–a.e. x and that*

$$\int_{\mathcal{X}} \left(\|g(x)\|_H^2 + \|\tilde{D}g(x)\|_{H^{\otimes 2}}^2 \right) \omega(x) \mu(dx) < \infty.$$

Then $g \in D(L)$. Furthermore, there exists a positive, measurable function $k : \mathcal{X} \to \mathbb{R}$, depending only on ω, such that

$$\int_{\mathcal{X}} |Lg(x)|^2 k(x) \mu(dx) \le \int_{\mathcal{X}} \left(\|g(x)\|_H^2 + \|\tilde{D}g(x)\|_{H^{\otimes 2}}^2 \right) \omega(x) \mu(dx) < \infty.$$

Proof. The proof in [70] applies, if instead of references to Theorem 5.1 and to Corollary to Theorem 4.2 [70], references to Theorem 6.5.8 and Corollary 6.5.3 are made.   □

Theorem 6.5.10. $H - C^1 \subseteq \mathcal{D}(L)$.

Proof. Clearly $\omega(x) = \{1 + \|g(x)\|_H^2 + \|D^{\mathcal{F}}g(x)\|_{H^{\otimes 2}}^2\}^{-1}$ is a weight function for $g \in H - C^1$.

Also $H - C^1 \subset H^1(\mathcal{X} \to H, d\mu)$. Indeed, first, $g \in H - C^1$ implies Fréchet differentiability of g, which is stronger than the SGD property. Also, g is Bochner measurable because it is measurable and H is separable.

Further, we need the Bochner measurability of the H-Fréchet derivative of g, $D^{\mathcal{F}} g : \mathcal{X} \to L(H)$. We have for all $x \in \mathcal{X}$ and $k \in H$,

$$\frac{1}{t} \langle (g(x+th) - g(x)), k \rangle_H \to (D^{\mathcal{F}} g(x))(h \otimes k).$$

The LHS of the above expression is measurable. Therefore, the RHS, as a limit, is measurable. Thus, $D^{\mathcal{F}} g : \mathcal{X} \to H^{\otimes 2}$ is weakly measurable. But $H^{\otimes 2}$ is a separable Hilbert space, therefore $D^{\mathcal{F}} g$ is Bochner measurable as a map from $\mathcal{X}$ to $L(H)$ because $H^{\otimes 2} \hookrightarrow L(H)$ continuously. The RAC condition for g follows from the inequality

$$\sum_{i=1}^{n} \|g(x+t_{i+1}h) - g(x+t_i h)\|_H$$

$$\leq \sup_{\alpha \in [a,b]} \|D^{\mathcal{F}} g(x + \alpha h)\|_{H^{\otimes 2}} \|h\|_H (b-a)$$

where $a = t_1 \leq t_2 \leq \ldots \leq t_{n+1} = b$ is a partition of an interval $[a,b]$ (see [70], p. 570). $\qquad\square$

Lemma 6.5.4. *Assume that $F : \mathcal{X} \to H$ is in class $H - C^1$ and is an H-contraction, then there exists a function $G \in \mathcal{D}(L)$, such that*

$$(I_{\mathcal{X}} - F)^{-1} = (I_{\mathcal{X}} - G)$$

and

$$\frac{d\mu \circ (I_{\mathcal{X}} - F)^{-1}}{d\mu} = |d(x,G)|.$$

Proof. By Lemma 6.5.1 $(I_{\mathcal{X}} - F) : \mathcal{X} \to \mathcal{X}$ is bijective. Define

$$G(x) = x - (I_{\mathcal{X}} - F)^{-1}(x).$$

Then

$$(I_{\mathcal{X}} - F)^{-1} = (I_{\mathcal{X}} - G), \tag{6.26}$$

and (see Exercise 6.5.5)

$$G(x) = -F\left((I_{\mathcal{X}} - F)^{-1}\right)(x). \tag{6.27}$$

The implicit function theorem, Theorem 1.20 and Corollary 1.21 in [113], applied to (6.27), guarantees that $G : \mathcal{X} \to H$ is an $H - C^1$ map since the derivative is a composition of a Hilbert–Schmidt operator and linear operators. Also, using (6.27), for all $x \in \mathcal{X}$

$$G((I_\mathcal{X} - F)(x)) + F(x) = -F\left((I_\mathcal{X} - F)^{-1}(I_\mathcal{X} - F)(x)\right) + F(x) = 0. \quad (6.28)$$

By differentiating,

$$\begin{aligned} D^{\mathcal{F}}\left(G((I_\mathcal{X} - F)(x)) + F(x)\right) & \quad (6.29) \\ = D^{\mathcal{F}} G((I_\mathcal{X} - F)(x))\left(I_H - D^{\mathcal{F}} F(x)\right) + DF(x) = 0, \end{aligned}$$

implying that (Exercise 6.5.5)

$$I_H = \left(I_H - D^{\mathcal{F}} G((I_\mathcal{X} - F)(x))\right)\left(I_H - D^{\mathcal{F}} Fx\right), \quad (6.30)$$

and

$$I_H - D^{\mathcal{F}} G(x) = \left(I_H - D^{\mathcal{F}} F\left((I_\mathcal{X} - F)^{-1}\right)\right)^{-1}. \quad (6.31)$$

Knowing that $F, G \in H - C^1 \subseteq \mathcal{D}(L)$, we can perform the following calculations, here $P \in \mathcal{P}(H)$. Using (6.28),

$$\begin{aligned} 0 & = L_P(0) = L_P\left(G((I_\mathcal{X} - F)) + F\right)(x) \\ & = \langle PG(I_\mathcal{X} - F)(x) + PF(x), x \rangle - \mathrm{tr} PD^{\mathcal{F}}\left(G(I_\mathcal{X} - F)(x) + F(x)\right) \\ & = \langle PG(I_\mathcal{X} - F)(x), x \rangle + \langle PF(x), x \rangle \\ & \quad - \mathrm{tr} PD^{\mathcal{F}}\left(G(I_\mathcal{X} - F)(x)\right) - \mathrm{tr} PD^{\mathcal{F}} F(x) \\ & = \langle PG(I_\mathcal{X} - F)(x), x - F(x) \rangle - \mathrm{tr} P(D^{\mathcal{F}} G)((I_\mathcal{X} - F)(x)) \\ & \quad + \langle PF(x), x \rangle - \mathrm{tr} P(D^{\mathcal{F}} F)(x) + \langle PG((I_\mathcal{X} - F)(x)), F(x) \rangle_H \\ & \quad + \mathrm{tr}\left((PD^{\mathcal{F}} G)((I_\mathcal{X} - F)(x)) \circ D^{\mathcal{F}} F(x)\right) \\ & = L_P G((I_\mathcal{X} - F)(x)) + L_P F(x) + \langle PG((I_\mathcal{X} - F)(x)), F(x) \rangle_H \\ & \quad + \mathrm{tr}\left(P(D^{\mathcal{F}} G)((I_\mathcal{X} - F)(x)) \circ D^{\mathcal{F}} F(x)\right) \\ & \rightarrow LG((I_\mathcal{X} - F)(x)) + LF(x) + \langle G((I_\mathcal{X} - F)(x)), F(x) \rangle_H \\ & \quad + \mathrm{tr}\left((D^{\mathcal{F}} G)((I_\mathcal{X} - F)(x)) \circ D^{\mathcal{F}} Fx\right), \end{aligned}$$

in probability μ, since $(I_\mathcal{X} - F)\mu \ll \mu$ by Corollary 6.5.2. We have shown that

$$\begin{aligned} L\left(G((I_\mathcal{X} - F)(x)) + F(x)\right) = LG((I_\mathcal{X} - F)(x)) + LF(x) \\ + \langle G((I_\mathcal{X} - F)(x)), F(x) \rangle_H + \mathrm{tr}\left((D^{\mathcal{F}} G)((I_\mathcal{X} - F)(x)) \circ D^{\mathcal{F}} F(x)\right) = 0 \end{aligned}$$

Using this fact and (6.28), we now calculate

$$\begin{aligned} 1 & = d(x, 0) = d\left(x, G((I_\mathcal{X} - F)(x)) + F(x)\right) \\ & = d_c\left(I_H - D^{\mathcal{F}}\left(G((I_\mathcal{X} - F)(x)) + F(x)\right)\right) \end{aligned}$$

$$\times \exp\left(L(G((I_{\mathcal{X}} - F)(x)) + F(x)) - \frac{1}{2}\|G((I_{\mathcal{X}} - F)(x)) + F(x)\|_H^2 \right)$$

$$= \exp\left(LG((I_{\mathcal{X}} - F)(x)) - \frac{1}{2}\|G((I_{\mathcal{X}} - F)(x))\|_H^2 \right)$$

$$\times \exp\left(LF(x) - \frac{1}{2}\|F(x)\|_H^2 \right) \exp\left(\mathrm{tr}\left(D^{\mathcal{F}} G((I_{\mathcal{X}} - F)(x)) D^{\mathcal{F}} F(x) \right) \right)$$

$$= d_c\left(I_H - D^{\mathcal{F}} G((I_{\mathcal{X}} - F)(x)) \right)$$

$$\times \exp\left(LG((I_{\mathcal{X}} - F)(x)) - \frac{1}{2}\|G((I_{\mathcal{X}} - F)(x))\|_H^2 \right)$$

$$\times d_c\left(I_H - D^{\mathcal{F}} F(x) \right) \exp\left(LF(x) - \frac{1}{2}\|F(x)\|_H^2 \right)$$

$$\times \exp\left(\mathrm{tr}\left(D^{\mathcal{F}} G((I_{\mathcal{X}} - F)(x)) D^{\mathcal{F}} F(x) \right) \right),$$

where in the last step we have used property (e) in Exercise 6.5.1 of Carleman–Fredholm determinant applied to the identity operator (recall (6.30)) $\left(I_H - D^{\mathcal{F}} G((I_{\mathcal{X}} - F)(x)) \right)\left(I_H - D^{\mathcal{F}} F(x) \right)$. We have shown that

$$d((I_{\mathcal{X}} - F)(x), G) d(x, F) = 1, \quad \mu\text{–a.e.} \tag{6.32}$$

$\square$

Exercise 6.5.5. *Show* (6.27), (6.30), *and* (6.31).

In the next lemma we utilize the Novikov condition (assumption (b)) to ensure uniform integrability of the sequence of densities in the Girsanov theorem.

Lemma 6.5.5. *Let F, $\{F_n\}_{n=1}^{\infty}$ be Bochner measurable transformations on $\mathcal{X}$ to H such that $F_n \to F$ pointwise. Assume that either*

(a) *For all n, F_n is an H-contraction and $\sup_n \left\| \tilde{D} F_n(x) \right\|_{H^{\otimes 2}} \le c < 1$ μ–a.e.*
or

(a') *$\sup_n \left\| D^{\mathcal{F}} F_n(x) \right\|_{H^{\otimes 2}} \le c < 1$, $\forall x \in \mathcal{X}$.*
Moreover, let

(b) *$\sup_n E_\mu\left\{ e^{\frac{q}{2}\|F_n\|_H^2} \right\} < \infty$ for some $q > 1$.*
Then for some $p > 1$,

$$\sup_n \left(\int_{\mathcal{X}} |d(x, F_n)|^p \mu(dx) \right)^{\frac{1}{p}} < \infty.$$

Proof. Let $\varepsilon > 0$ satisfy $c < 1 - \varepsilon$ and $(1 - \varepsilon)q > 1$. Choose $p' = 1 + \frac{\varepsilon}{(1-\varepsilon)^2 q}$ and $p \in (1, \frac{(1-\varepsilon)q}{(1-\varepsilon)^2 q + \varepsilon})$. Then

$$E_\mu |d(x, F_n)|^p = E\{ |d_c(I_H - \tilde{D} F_n(x))|^p \exp\{ pLF_n(x) - \frac{p}{2}\|F_n(x)\|_H^2 \} \}. \tag{6.33}$$

But by [116], for some $\Gamma > 0$

$$|d_c(I_H - \tilde{D}F_n(x))|^p \leq \exp\{p\Gamma\|\tilde{D}F_n(x)\|^2_{H\otimes 2}\} \leq \exp(p\Gamma c^2) = C_0$$

giving

$$E_\mu |d(x, F_n)|^p \leq C_0 E_\mu \left\{ \exp\{pLF_n(x) - \frac{p}{2}\|F_n\|^2_H\} \right\}.$$

By Hölder's inequality, the last expression is bounded by

$$C_0 E_\mu \left\{ \exp\{pp'LF_n(x) - \frac{(pp')^2}{2}\|F_n\|^2_H\} \right\}^{\frac{1}{p'}} E_\mu \left\{ \exp\{\frac{pp'(pp'-1)}{2(p'-1)}\|F_n\|^2_H\} \right\}^{\frac{p'-1}{p'}}.$$

Note $\frac{p'p(pp'-1)}{p'-1} < q$. With $\bar{F}_n = pp'F_n$, we get

$$\begin{aligned}
E_\mu |d(x, F_n)|^p &\leq C_0 E_\mu \left\{ \exp\{L\bar{F}_n(x) - \frac{1}{2}\|\bar{F}_n(x)\|^2_H\} \right\}^{\frac{1}{p'}} \\
&\quad \times E_\mu \left\{ \exp\{\frac{q}{2}\|F_n(x)\|^2_H\} \right\}^{\frac{p'-1}{p'}}.
\end{aligned}$$

In view of condition (b), it suffices to bound the first factor which equals

$$E_\mu \left\{ |d(x, \bar{F}_n)| \frac{1}{|d_c(I_H - \tilde{D}\bar{F}_n(x))|} \right\}^{\frac{1}{p'}}.$$

Now, observe that $pp'c < 1$. Therefore, as $(1+x)e^{-x} \geq e^{\Gamma'x^2}$ for $x < c < 1$ and some constant $\Gamma' < 0$, we get that the above expression does not exceed

$$E_\mu \left\{ |d(x, \bar{F}_n)| \sup_x e^{-\Gamma'\|\tilde{D}\bar{F}_n(x)\|_{H\otimes 2}} \right\}^{\frac{1}{p'}}$$

which is bounded by condition (a) and the fact that $|d(x, \bar{F}_n)|$ are densities (see (6.20) of Corollary 6.5.2).   □

Using Lemmas 6.5.1 and 6.5.5 we arrive at the following theorem:

Theorem 6.5.11. *Let F be a Bochner measurable transformation with $F \in \mathcal{D}(L)$, F an H-contraction, and $\|\tilde{D}F\|_{H\otimes 2} \leq c < 1$ μ–a.e. Suppose that for some $q > 1$,*

$$\sup_n E_\mu \left\{ e^{\frac{q}{2}\|F_n\|^2_H} \right\} < \infty$$

and $T = (I_X - F)$. Then,

(a) *T is bijective.*

(b) $\mu \sim \mu \circ T$ and $\frac{d\mu \circ T}{d\mu} = |d(x,F)|$

(c) $\frac{d\mu \circ T^{-1}}{d\mu}(T(x)) = |d(x,F)|^{-1}$

Note that since $\sup_n \|F_n\|_H^2 = \|F\|_H^2$, Theorem 6.5.11 gives conditions on F.

Example 6.5.1 (Wiener Space). *Consider* $\mathcal{X} = C[0,1]$ *and* μ, *the Wiener measure. Let* $F(x) = \int_0^\cdot K_s(x)ds$ *where* $K_s(x) \in L_2(\mathcal{X}, L_2([0,1]))$ *and assume that the Malliavin (Skorokhod) derivative* $D^M K$ *satisfies* $(\int_0^1 \int_0^1 (D_t^M K_s(x))^2 dsdt)^{1/2} < 1$ *a.e.* $[\mu]$. *Then one can approximate* F *by a sequence* $\{\tilde{F}_n\}_{n=1}^\infty$ *given by "smooth step processes" of the form*

$$K_s^n(x) = \sum_{j=1}^n f_j(\langle 1_{\Delta_1}, x\rangle, ..., \langle 1_{\Delta_n}, x\rangle) 1_{\Delta_j}(s),$$

($\langle 1_\Delta, x\rangle$ is the increment of x *on* Δ), *such that*

$$\sup_n \sup_{x \in \mathcal{X}} \|\tilde{D}\tilde{F}_n(x)\|_{H^{\otimes 2}}^2 = \sup_n \sup_{x \in \mathcal{X}} \|D^M \tilde{F}_n(x)\|_{H^{\otimes 2}}^2 < 1,$$

(see Proposition 2.6 in [9]). It is easy to check (see Lemma 3.2 in [9]) that the functions $\tilde{F}_n$ *are H-contractions and hence we obtain by Theorem 6.5.5, equalities (6.19) and (6.20) for transformations* $I_\mathcal{X} - \tilde{F}_n$ *(the same conclusion can be obtained by appealing to Theorem 6.4 in [70]).*

Because $I_\mathcal{X} - \tilde{F}_n \xrightarrow{\mu} I_\mathcal{X} - F$, *we obtain*

$$d(x, \tilde{F}_n) \xrightarrow{\mu} d(x, F),$$

under the assumption that $F \in \mathcal{D}(\tilde{D})$, *with* $D^M F(x) = \tilde{D}F(x)$, *a.e.* $[\mu]$. *For this it suffices only to require that* F *satisfies the following (G-S) condition (see [31]): for any* $k, k' \in H$

$$\frac{1}{\varepsilon}(\langle F(x+\varepsilon k), k'\rangle_H - \langle F(x), k'\rangle_K) \qquad \text{(G-S)}$$

converges in $L_2(\mathcal{X})$ *as* $\varepsilon \to 0$.

Moreover, by Lemma 4.2 in [9], if $E_\mu\{e^{\frac{q}{2}\|F\|_H^2}\} < \infty$ *then the approximating sequence can be chosen to satisfy*

$$\sup_n E_\mu\{e^{\frac{q}{2}\|\tilde{F}_n\|_H^2}\} < \infty$$

hence, $\{d(x, \tilde{F}_n)\}_{n=1}^\infty$ *is uniformly integrable by Lemma 6.5.5, and we obtain (6.17). Also, the bound on the Hilbert-Schmidt norm of* $D^M K$ *guarantees the existence of a transformation* $A : \mathcal{X} \to \mathcal{X}$ *with* $A \circ T = T \circ A = I_\mathcal{X}$, μ–*a.e., so that we can obtain (6.19) and (6.20) (one can also use Theorem 6.4 in [70]). The following result in [9] follows.*

Theorem 6.5.12 (Bukhdan). *Let $\mathcal{X} = C[0,1]$ and μ be the Wiener measure. Consider $K_s(x) \in L_2(\mathcal{X}, L_2([0,1]))$ with the Malliavin derivative $D^M K$ satisfying*

$$\left(\int_0^1 \int_0^1 (D_t^M K_s(x))^2 ds dt \right)^{1/2} < 1 \; \mu - a.e.$$

and

$$E_\mu \left\{ e^{\frac{q}{2} \int_0^1 K_s^2 ds} \right\} < \infty.$$

Then, for $T : \mathcal{X} \to \mathcal{X}$ given by

$$T(x) = x - \int_0^{\cdot} K_s(x) ds$$

we have $\mu \circ T \ll \mu$. If, moreover, $F(x) = \int_0^{\cdot} K_s(x) ds$ satisfies condition (G-S), then the density is given by $|d(x, F)|$.

We now wish to find an upper bound on $E_\mu |d(x, F_n)|^p$. This will allow us to drop the H-contraction condition on F. However, we need a condition stronger than (b) of Theorem 6.5.5, which is difficult to verify. To secure uniform integrability from our calculations, we need the following lemma in [54].

Lemma 6.5.6 (Kallianpur and Karandikar). *Let $A \in H^{\otimes 2}$ be such that $I_H + tA$ is invertible for $t \in \mathbf{R}$ with $|1-t| < \eta, \eta > 0$. Let $1 < s < 1 + \frac{\eta}{2}, s < 2$. Then there exists a constant $C = C(s, \eta)$, so that*

$$|d_c(I_H + A)|^s |d_c(I_H + sA)|^{-1} \le \exp\{C \|A\|_{H^{\otimes 2}}^2\}.$$

The following theorem is an analogue of Theorem 6 of Kallianpur and Karandikar in [54] for our case.

Theorem 6.5.13. *Let $F \in \mathcal{D}(L)$ satisfy the following conditions:*

(a) *For all $1 \le a < 1 + \eta$, $n = 1, 2, \ldots$, aF_n satisfies condition (b) of Theorem 1.*

(b) *$\sup_n E_\mu \{ e^{\frac{q}{2} \|F_n\|_H^2} \} < \infty$ for some $q > 1$.*

(c) *$I_H - t\tilde{D}F_n$ is invertible with $|1-t| < \eta$, where $\eta > 0$.*

Then, assumptions of Theorem 6.5.5 are satisfied, equality (6.17) holds true, and under the additional assumption that $I_{\mathcal{X}} - F$ is bijective, quality (6.20).

Proof. Let us compute

$E_\mu |d(x, F_n)|^p$

$$= E_\mu \left\{ |d_c(I_H - \tilde{D}F_n)^p \exp\{pLF_n - \frac{p}{2} \|F_n\|_H^2\} \right\}$$

$$= E_\mu \left\{ |d_c(I_H - \tilde{D}F_n)|^p \exp\{pLF_n - \frac{p^2 p'}{2} \|F_n\|_H^2\} \exp\{\frac{p^2 p'}{2} \|F_n\|_H^2 - \frac{p}{2} \|F_n\|_H^2\} \right\}$$

$$= E_\mu \left\{ \frac{|d_c(I_H - \tilde{D}F_n)|^p}{|d_c(I_H - pp'\tilde{D}F_n)|^{\frac{1}{p'}}} \left[|d_c(I_H - pp'\tilde{D}F_n)|^{\frac{1}{p'}} \exp\{pLF_n - \frac{p^2 p'}{2} \|F_n\|_H^2\} \right] \right.$$

$$\times \exp\{(\tfrac{p^2 p'}{2} - \tfrac{p}{2}) \|F_n\|_H^2\}\}.$$

By Hölder's inequality

$$E_\mu |d(x, F_n)|^p$$

$$\le \left[E_\mu \left\{ \frac{|d_c(I_H - \tilde{D}F_n)|^p}{|d_c(I_H - pp'\tilde{D}F_n)|^{\frac{1}{p'}}} \exp\{(\tfrac{p^2 p'}{2} - \tfrac{p}{2}) \|F_n\|_H^2\} \right\}^{\frac{p'}{p'-1}} \right]^{\frac{p'-1}{p'}}$$

$$\times \left[E_\mu \left\{ |d_c(I_H - pp'\tilde{D}F_n)| \exp\{L(pp'F_n) - \tfrac{1}{2}\|pp'F_n\|_H^2\} \right\} \right]^{\frac{1}{p'}}$$

$$= \left[E_\mu \left\{ \left(\frac{|d_c(I_H - \tilde{D}F_n)|^{pp'}}{|d_c(I_H - pp'\tilde{D}F_n)|} \right)^{\frac{1}{p'-1}} \exp\{ \frac{pp'(pp'-1)}{2(p'-1)} \|F_n\|_H^2 \} \right\} \right]^{\frac{p'-1}{p'}} \cdot B_n^{\frac{1}{p'}},$$

with B_n denoting the second expectation in the product above. Now, there exists $r > 1$, such that $\frac{pp'(pp'-1)}{(p'-1)} r < q$, the Novikov's constant of condition (b) of Lemma 6.5.5. Choose such an r and let $r' = \frac{r}{r-1}$. Then the last expectation, by Hölder's inequality, is bounded above by

$$\left[E_\mu \left\{ \left(\frac{|d_c(I_H - \tilde{D}F_n)|^{pp'}}{|d_c(I_H - pp'\tilde{D}F_n)|} \right)^{\frac{r'}{p'-1}} \right\} \right]^{\frac{1}{r'}} \left[E_\mu \left\{ \exp\{\tfrac{q}{2}\|F_n\|_H^2\} \right\} \right]^{\frac{1}{r}}.$$

If we assume that for $1 \le a < 1 + \eta$, aF_n satisfies condition (b) of Theorem 6.5.5, then by the first part of the proof of Theorem 6.5.5, we can conclude that $|d(x, pp'F_n)|$ is a density for $pp' < 1 + \eta$, giving $B_n = 1$ for all n. Condition (b) of Lemma 6.5.5 gives that the second expectation in the above inequality is bounded. By Lemma 6.5.6, under the assumption that $I_H - t\tilde{D}F_n$ is invertible with t as in the lemma, we get that the first expectation is bounded by

$$\sup_n E_\mu \exp\left\{ C(pp', \eta) \cdot \frac{r'}{p'-1} \|\tilde{D}F_n(x)\|_{H^{\otimes 2}}^2 \right\} < \infty,$$

if we choose $pp' < 1 + \tfrac{\eta}{2}$. $\square$

To obtain an analogue of the result of Enchev [23], Theorem 3, we assume that $F : \mathcal{X} \to H$, $F \in \mathcal{D}(L)$ and F is an H-contraction (thus we do not require any bound on the Hilbert-Schmidt norm of $\tilde{D}F$). Because for $P_n \to I_{\mathcal{X}}$, $P_n \in P^*(H)$ we have,

$$L_{P_n} F \overset{\mu}{\to} LF, \qquad \|F_n\|_H \to \|F\|_H \qquad \text{and} \quad \|\tilde{D}F_n - \tilde{D}F\|_{H^{\otimes 2}} \to 0$$

we have

$$d(x, F_n) \overset{\mu}{\to} d(x, F). \tag{6.34}$$

Now, in a similar way as we obtained (6.19) for the transformation $T = I_{\mathcal{X}} - F$, we can derive its analogue for the transformation $T_n = I_{\mathcal{X}} - F_n$ from relation (6.18),

$$\int_{\mathcal{X}} g(x) \mu \circ (I_{\mathcal{X}} - F_n)(dx) = \int g(x) |d(x, F_n)| \mu(dx)$$

which gives $|d(x, F_n)| = \frac{d\mu \circ (I_{\mathcal{X}} - F_n)}{d\mu}(x)$.

Enchev's assumption $E_\mu |d(x, F)| = 1$ together with the convergence in (6.34) guarantees uniform integrability of $\{d(x, F_n)\}_{n=1}^{\infty}$ and hence, we obtain (6.5.5). From this it follows that

$$E_\mu \{ e^{i(\ell, (I_{\mathcal{X}} - F)(x))} |d(x, F)| \} = e^{-\frac{1}{2} \|\ell\|_H^2} \tag{6.35}$$

for $\ell \in \mathcal{X}^* \subset H$. That is the distribution of the element $(I_{\mathcal{X}} - F)(x)$ under the measure $|d(x, F)| \mu(dx)$ is the isonormal measure on $\mathcal{X}$. Thus, we obtain a result that if $F \in \mathcal{D}(L)$ is an H-contraction and $E_\mu |d(x, F)| = 1$, then (6.35) holds.

In the particular case, if $F(x)$ is the Skorohod integral of a function $\varphi \in L_2(\mathcal{X}, H)$, we obtain the result of Enchev. And if φ is non-anticipative, we obtain Girsanov's result in view of the fact that the Carleman-Fredholm determinant is equal to 1 (see [123]).

Remark 6.5.2. *If $\mathcal{X} = C[0, 1]$ and μ is the Wiener measure, one can approximate F by F_n as in Remark 6.5.1. Under the condition $\|DF(x)\|_{H^{\otimes 2}} \le c < 1$, for all $x \in \mathcal{X}$, following the arguments of Remark 6.5.1, transformations F_n can be chosen to be H-contractions such that the convergence in (6.34) holds. In this case, assuming that $E_\mu |d(x, F)| = 1$, again we obtain Enchev's result (6.35).*

Chapter 7

Markov Property of Gaussian Fields

7.1 Linear Functionals on the Space of Radon Signed Measures

Let E be a separable locally compact Hausdorff space. In this section we introduce the space $M(E)$, consisting of Radon signed measures on $\mathcal{B}(E)$, the Borel subsets of E, with compact support in E, and discuss properties of linear functionals on $M(E)$. We begin with basic definitions.

Definition 7.1.1. *Let E be a locally compact Hausdorff second-countable topological space (i.e., with a countable base for its topology). The support of a signed measure μ on E, denoted by* $\mathrm{supp}(\mu)$*, is defined to be the complement of the largest open set $\mathcal{O} \subseteq E$, such that $|\mu|(\mathcal{O}) = 0$, where $|\mu|$ is the total variation measure of μ, that is, $|\mu| = \mu^+ + \mu^-$, where μ^+ and μ^- are unique mutually singular measures on E, such that $\mu = \mu^+ - \mu^-$ (this decomposition of μ is called the Jordan decomposition, [110]). We define $M(E)$ as the set of Radon signed measures with compact support in E.*

For $f : M(E) \to \mathbb{R}$, we define its support, $\mathrm{supp}(f)$*, as the complement of the largest open set $\mathcal{O} \subseteq E$, such that $f(\mu) = 0$ for all $\mu \in M(E)$ with $\mathrm{supp}(\mu) \subseteq \mathcal{O}$.*

Definition 7.1.2. *We say that a measure $\mu \in M(E)$ has a partition of unity property if for any open covering $\{\mathcal{O}_1, ..., \mathcal{O}_n\}$ of the support of μ,* $\mathrm{supp}(\mu)$*, there exist measures $\mu_1, ..., \mu_n \in M(E)$ with $\mathrm{supp}(\mu_i) \subseteq \mathcal{O}_i$, $i = 1, 2, .., n$, and such that $\mu = \mu_1 + \mu_2 + .. + \mu_n$.*

We make the following assumptions about $M(E)$:

(A1) $M(E)$ is a real vector space,

(A2) $M(E)$ has *partition of unity property*,

(A3) If f is a linear functional on $M(E)$ and $\mathrm{supp}(f) \subseteq A_1 \cup A_2$, where A_1, A_2 are two disjoint closed subsets of E, then $f = f_1 + f_2$, where f_1, f_2 are linear functionals on $M(E)$ with $\mathrm{supp}(f_i) \subseteq A_i$, $i = 1, 2$.

We show that under assumptions (A1) and (A2) the support of a linear functional on $M(E)$ is well defined and introduce its fundamental properties.

Lemma 7.1.1. *Assume (A1) and (A2), then*

(a) *If f is a linear functional on $M(E)$, then $\mathrm{supp}(f) = \left(\bigcup_i \mathcal{O}_i \right)^c$ where the union is taken over all open sets $\mathcal{O}_i$, such that $f(\mu) = 0$ for all $\mu \in M(E)$ with $\mathrm{supp}(\mu) \subseteq \mathcal{O}_i$,*

(b) *If f is a linear functional on $M(E)$ and $\operatorname{supp}(f)$ is an empty set, then $f = 0$, that is, $f(\mu) = 0$ for all $\mu \in M(E)$,*

(c) *f_1, f_2 are linear functionals on $M(E)$, then $\operatorname{supp}(f_1 + f_2) \subseteq \operatorname{supp}(f_1) \cup \operatorname{supp}(f_2)$.*

Proof. To prove (a) note that if $\mathcal{O} \subseteq E$ is an open set such that $f(\mu) = 0$ for all $\mu \in M(E)$ with $\operatorname{supp}(\mu) \subseteq \mathcal{O}$, then $\mathcal{O}$ is included in the union. We only need to show that the union $\bigcup_i \mathcal{O}_i$ itself has the property that $f(\mu) = 0$ whenever $\operatorname{supp}(\mu) \subseteq \bigcup_i \mathcal{O}_i$. By the compactness of the support, for any such measure μ there exists a finite covering $\mathcal{O}_{i_1}, ..., \mathcal{O}_{i_n}$ of $\operatorname{supp}(\mu)$. Using property (A2), we conclude that $\mu = \mu_1 + ... + \mu_n$, $\mu_i \in M(E)$, with $\operatorname{supp}(\mu_i) \subseteq \mathcal{O}_i$, $i = 1, ..., n$, giving that

$$f(\mu) = f(\mu_1 + ... + \mu_n) = f(\mu_1) + ... + f(\mu_n) = 0$$

by the definition of the sets $\mathcal{O}_i$.

Part (b) follows from the definition of the support of f.

To prove (c), denote $A_i = \operatorname{supp}(f_i)$, $i = 1, 2$. Let $\mu \in M(E)$ with $\operatorname{supp}(\mu) \subseteq (A_1 \cup A_2)^c = A_1^c \cap A_2^c$. Then $\operatorname{supp}(\mu) \subseteq A_i^c$, $i = 1, 2$, giving $f_i(\mu) = 0$, $i = 1, 2$, and $f_1(\mu) + f_2(\mu) = 0$. Hence, $\operatorname{supp}(f_1 + f_2) \subseteq A_1 \cup A_2$. $\qquad\square$

Remark 7.1.1. *If E is a normal space and $\{\mathcal{O}_1, \mathcal{O}_2, ..., \mathcal{O}_n\}$ is an open covering of a closed set A, then there exist open sets $U_1, U_2, ..., U_n$ such that $\overline{U}_i \subseteq \mathcal{O}_i$, $i = 1, ..., n$, and $\bigcup_{i=1}^{n} U_i \supseteq A$ ($\overline{U}_i$ denotes the closure of U_i). In our case, E is a normal space.*

Exercise 7.1.1. *Prove Remark 7.1.1 using induction.*

We give an alternate form for condition (A3).

Lemma 7.1.2. *Under assumptions (A1) and (A2), conditions (A3) and (A3') below are equivalent.*

(A3') *If f is a linear functional on $M(E)$ and $\operatorname{supp}(f) \subseteq A_1 \cup A_2$, with A_1, A_2 disjoint closed sets, then for any two disjoint open sets $\mathcal{O}_1, \mathcal{O}_2$, with $A_i \subseteq \mathcal{O}_i$, $i = 1, 2$, there exist linear functionals f_1, f_2 on $M(E)$, with $\operatorname{supp}(f_i) \subseteq \mathcal{O}_i$, $i = 1, 2$, such that $f = f_1 + f_2$.*

Proof. It is obvious that condition (A3) implies (A3'). To prove the converse, let $\mathcal{O}_1, \mathcal{O}_2$ be two disjoint open subsets of E, such that $\mathcal{O}_i \supseteq A_i$, $i = 1, 2$. Then $f = f_1 + f_2$, where f_i are linear functionals on $M(E)$ with $\operatorname{supp}(f_i) \subseteq \mathcal{O}_i$, $i = 1, 2$. Now take another pair of open sets $\mathcal{O}_i' \subseteq \mathcal{O}_i$, such that $A_i \subseteq \mathcal{O}_i'$, $i = 1, 2$. Then $f = f_1' + f_2'$ with $\operatorname{supp}(f_i') \subseteq \mathcal{O}_i'$, $i = 1, 2$, so that $f_1 - f_1' = f_2 - f_2'$. By Lemma 7.1.2(c), $\operatorname{supp}(f_1 - f_1') \subseteq \operatorname{supp}(f_1) \cup \operatorname{supp}(f_1') \subseteq \mathcal{O}_1$ and $\operatorname{supp}(f_2 - f_2') \subseteq \operatorname{supp}(f_2) \cup \operatorname{supp}(f_2') \subseteq \mathcal{O}_2$. Since $\mathcal{O}_1 \cap \mathcal{O}_2 = \varnothing$ we conclude that $f_1 - f_1' = f_2 - f_2' = 0$, which implies that $\operatorname{supp}(f_1) \subseteq \mathcal{O}_1'$. It follows by induction that

$$\text{supp}(f_1) \subseteq \bigcap_{A_1 \subseteq \mathcal{O} \subseteq \mathcal{O}_1} \mathcal{O} = A_1$$

and similarly, $\text{supp}(f_2) \subseteq A_2$.   □

Example 7.1.1 (Infinitely differentiable functions). *Let E be an open domain in $\mathbb{R}^d$, and*

$$M(E) = \{\mu \mid d\mu = \varphi\, dx,\ \varphi \in C_0^\infty(E)\},$$

where $C_0^\infty(E)$ consists of all infinitely differentiable real-valued functions with compact support in E and E is equipped with the relative topology. Clearly, $M(E)$ is a real vector space. If $\varphi \in C_0^\infty(E)$ and $\{\mathcal{O}_1, \mathcal{O}_2, ..., \mathcal{O}_n\}$ is an open covering of $\text{supp}(\mu)$, then there exist (by the existence of smooth partition of unity) functions $\varphi_1, \varphi_2, ..., \varphi_n \in C_0^\infty(E)$ with $\text{supp}(\varphi_i) \subseteq \mathcal{O}_i$, $i = 1, ..., n$ and $\varphi = \sum_{i=1}^n \varphi_i$. Here, $\text{supp}(\varphi) = \overline{\{x \in E \mid \varphi(x) \neq 0\}}$. Since $\text{supp}(\varphi) = \text{supp}(\varphi\, dx)$ we can see that condition (A2) holds true.

To verify condition (A3'), we allow f to be a linear functional on $M(E)$ such that $\text{supp}(f) \subseteq A_1 \cup A_2$, with A_1, A_2 disjoint closed subsets of E. For any two disjoint open sets $\mathcal{O}_1$, $\mathcal{O}_2$, such that $A_i \subseteq \mathcal{O}_i$, $(i = 1, 2)$, there exist two open sets $\mathcal{O}_i'$, $(i = 1, 2)$, with $A_i \subseteq \mathcal{O}_i' \subseteq \mathcal{O}_i$, $(i = 1, 2)$ and $\overline{\mathcal{O}_1'} \cap \overline{\mathcal{O}_2'} = \varnothing$. Let $\mathcal{O}_3' = (A_1 \cup A_2)^c$, then $\{\mathcal{O}_1', \mathcal{O}_2', \mathcal{O}_3'\}$ is an open covering of E. Then there exists $\varphi_i \in C_0^\infty(E)$, $i = 1, 2, 3$, such that $\varphi_i \geq 0$, $\text{supp}(\varphi_i) \subseteq \mathcal{O}_i'$ and $\sum_{i=1}^3 \varphi_i = 1$. Now for $\varphi \in C_0^\infty(E)$, $\varphi = \sum_{i=1}^3 \varphi\varphi_i$ and

$$f(\varphi) = f(\varphi\varphi_1) + f(\varphi\varphi_2) + f(\varphi\varphi_3).$$

Since $\text{supp}(\varphi\varphi_3) = \text{supp}(\varphi\varphi_3\, dx) \subseteq \mathcal{O}_3$ we have $f(\varphi\varphi_3) = 0$. Let $f_i(\varphi) = f(\varphi\varphi_i)$, $i = 1, 2$. If $\varphi \in C_0^\infty(E)$ with $\text{supp}(\varphi\, dx) = \text{supp}(\varphi) \subseteq (\text{supp}(\varphi_i))^c$, then, $\varphi\varphi_i = 0$, $i = 1, 2$. Thus, $f_i(\varphi) = 0$, $i = 1, 2$, if $\text{supp}(\varphi\, dx) \subseteq (\text{supp}(\varphi_i))^c$, $i = 1, 2$, implying that

$$\text{supp}(f_i) \subseteq ((\text{supp}(\varphi_i))^c)^c = \text{supp}(\varphi_i) \subseteq \mathcal{O}_i, \quad i = 1, 2$$

and giving (A3').

Example 7.1.2 (Measures of Finite Energy). *Let $M(E)$ be a vector space of Radon signed measures with compact support with the following property: if $\mu \in M(E)$, then $1_A\mu \in M(E)$ for all $A \in \mathcal{B}(E)$, the Borel subsets of E, where $1_A\mu(B) = \mu(A \cap B)$ for all $B \in \mathcal{B}(E)$. Then $M(E)$ satisfies conditions (A2) and (A3).*

To see this, suppose that $\mu \in M(E)$ with $\text{supp}(\mu) \subseteq \bigcup_{i=1}^n \mathcal{O}_i$, where $\mathcal{O}_i$, $i = 1, ..., n$, are open sets. By Remark 7.1.1, we can find open sets U_i, $(i = 1, 2, .., n)$, such that $\overline{U_i} \subseteq \mathcal{O}_i$ and $\text{supp}(\mu) \subseteq \bigcup_{i=1}^n U_i$. Since μ has compact support and is a finite measure, we have

$$\mu = \left(1_{\bigcup_{i=1}^n U_i}\right)\mu = 1_{U_1}\mu + 1_{U_2 \cap U_1^c}\mu + ... + 1_{U_n \cap U_1^c \cap ... \cap U_{n-1}^c}\mu$$

$$= \mu_1 + \mu_2 + \ldots + \mu_n.$$

Obviously, $\operatorname{supp}(\mu_i) \subseteq \overline{U_i} \subseteq \mathcal{O}_i$, $i = 1, \ldots, n$, *and* $\mu_i \in M(E)$ *for* $i = 1, \ldots, n$. *Hence condition (A2) holds.*

If f is a linear functional on $M(E)$ with $\operatorname{supp}(f) \subseteq A_1 \cup A_2$, *where* A_1, A_2 *are disjoint closed sets, then we choose open sets* $\mathcal{O}_1$, $\mathcal{O}_2$, *with* $A_i \subseteq \mathcal{O}_i$, $(i = 1, 2)$ *and* $\overline{\mathcal{O}_1} \cap \overline{\mathcal{O}_2} = \varnothing$. *Let* $\mathcal{O}_3 = (\mathcal{O}_1 \cup \mathcal{O}_2)^c$, *then* $E = \mathcal{O}_1 \cup \mathcal{O}_2 \cup \mathcal{O}_3$ *and the sets* $\mathcal{O}_i$, $i = 1, 2, 3$, *are disjoint. Then for* $\mu \in M(E)$,

$$\mu = 1_{\mathcal{O}_1}\mu + 1_{\mathcal{O}_2}\mu + 1_{\mathcal{O}_3}\mu = \mu_1 + \mu_2 + \mu_3.$$

Let $f_i(\mu) = f(\mu_i)$, $i = 1, 2, 3$, *so that*

$$f(\mu) = f(\mu_1) + f(\mu_2) + f(\mu_3) = f_1(\mu) + f_2(\mu) + f_3(\mu)$$

and f_1, f_2, f_3 *are linear functionals on* $M(E)$. *Since* $\operatorname{supp}(\mu_3) \subseteq \overline{\mathcal{O}_3} = \mathcal{O}_3 = (\mathcal{O}_1 \cup \mathcal{O}_2)^c \subseteq (A_1 \cup A_2)^c$ *we get* $f_3(\mu) = 0$ *for all* $\mu \in M(E)$. *Notice that for any* μ *with* $\operatorname{supp}(\mu) \subseteq \overline{\mathcal{O}_i}^c$, $i = 1, 2$, $\mu_i = 1_{\mathcal{O}_i}\mu \equiv 0$ *and hence* $f_i(\mu) = 0$, $i = 1, 2$ *for all* $\mu \in M(E)$ *with* $\operatorname{supp}(\mu) \subseteq \overline{\mathcal{O}_i}^c$. *Thus,* $\operatorname{supp}(f_i) \subseteq \overline{\mathcal{O}_i}$, $i = 1, 2$, *giving condition (A3').*

We shall now consider a random field $\{X_\mu, \mu \in M(E)\}$, with X being a linear function from $M(E)$ into the space of measurable functions $L^0(\Omega, \mathcal{F}, P)$. In case $\{X(t), t \in \mathbb{R}\}$ is a Markov process, then $\mathcal{F}_t^- = \sigma\{X(s), s \leq t\}$ (the past) and $\mathcal{F}_t^+ = \sigma\{X(s), s \geq t\}$ (the future) are conditionally independent given $\mathcal{G}_t = \sigma\{X(t)\}$ (the present).

Thus, in order to generalize the Markov property to random fields, it is tempting to say that X has Markov property on an open set $D \subseteq E$, if $\mathcal{F}_D^+ = \sigma\{X(\mu), \operatorname{supp}(\mu) \subseteq \overline{D}\}$ and $\mathcal{F}_D^- = \sigma\{X(\mu), \operatorname{supp}(\mu) \subseteq D^c\}$ are conditionally independent given $\mathcal{F}_{\partial D} = \sigma\{X(\mu), \operatorname{supp}(\mu) \subseteq \partial D\}$, where $\partial D = \overline{D} \cap D^c$ is the boundary of D. However, this notion is truly restrictive [87]. But clearly the Markov property is related to conditional independence. Before introducing the Markov property, we study some properties of conditional independence.

Let us start with the definition of conditional independence itself.

Definition 7.1.3. *Let $\mathcal{A}$, $\mathcal{B}$, and $\mathcal{G}$ be sub-σ-fields of a σ-field $\mathcal{F}$ in a probability space $(\Omega, \mathcal{F}, P)$. We say that $\mathcal{A}$ and $\mathcal{B}$ are conditionally independent given $\mathcal{G}$ if*

$$P(A \cap B|\mathcal{G}) = P(A|\mathcal{G})P(B|\mathcal{G}), \quad A \in \mathcal{A}, B \in \mathcal{B}, \tag{7.1}$$

where $P(\cdot|\mathcal{G})$ is conditional probability given $\mathcal{G}$. We denote this fact by $\mathcal{A} \perp\!\!\!\perp \mathcal{B}|\mathcal{G}$.

The following lemma provides a tool for verifying conditional independence.

Lemma 7.1.3. *Let $\mathcal{A}$, $\mathcal{B}$, $\mathcal{G}$ be sub-σ-fields of a σ-field $\mathcal{F}$, such that $\mathcal{G} \subseteq \mathcal{B}$. Then $\mathcal{A} \perp\!\!\!\perp \mathcal{B}|\mathcal{G}$ if and only if $E(f|\mathcal{G}) = E(f|\mathcal{B})$ for all bounded $\mathcal{A}$-measurable functions f. Here, $E(f|\mathcal{G})$ denotes conditional expectation of f given $\mathcal{G}$.*

Proof. Assume $E(f|\mathcal{G}) = E(f|\mathcal{B})$ for all bounded $\mathcal{A}$-measurable functions f, then for a bounded $\mathcal{B}$-measurable function g,

$$E(fg|\mathcal{G}) = E(E(fg|\mathcal{B})|\mathcal{G}) = E(gE(f|\mathcal{B})|\mathcal{G}) = E(g|\mathcal{G})E(f|\mathcal{G})$$

giving that $\mathcal{A} \perp\!\!\!\perp \mathcal{B}|\mathcal{G}$. To prove the converse, let $B \in \mathcal{B}$ and f be a bounded $\mathcal{A}$-measurable function. Then

$$\int_B f \, dP \;=\; \int_\Omega E(1_B f|\mathcal{G}) \, dP = \int_\Omega E(1_B|\mathcal{G}) E(f|\mathcal{G}) \, dP$$
$$=\; \int_B E(f|\mathcal{G}) \, dP,$$

where the second equality uses conditional independence and the fact that $E(1_B|\mathcal{G})E(f|\mathcal{G}) = E(1_B E(f|\mathcal{G})|\mathcal{G})$. $\qquad\square$

Definition 7.1.4. *We define $\mathcal{B} \vee \mathcal{G} = \sigma\{\mathcal{B} \cup \mathcal{G}\}$.*

Exercise 7.1.2. *Show that $\mathcal{B} \vee \mathcal{G} = \sigma\{B \cap A : B \in \mathcal{B}, A \in \mathcal{G}\}$.*

Use Exercise 7.1.2 and arguments as in the proof of Lemma 7.1.3 to do the following exercises.

Exercise 7.1.3. (a) $\mathcal{A} \perp\!\!\!\perp \mathcal{B}|\mathcal{G}$ *implies that $E(f|\mathcal{B} \vee \mathcal{G}) = E(f|\mathcal{G})$ for every bounded, $\mathcal{A}$-measurable function f.*

(b) $\mathcal{A} \perp\!\!\!\perp \mathcal{B}$ *implies that*

    (i) *for every $\tilde{\mathcal{G}}$ satisfying $\mathcal{G} \subseteq \tilde{\mathcal{G}} \subseteq \mathcal{G} \vee \mathcal{B}$, $\mathcal{A} \perp\!\!\!\perp \mathcal{B}|\tilde{\mathcal{G}}$,*

    (ii) *for every $\tilde{\mathcal{G}}$ satisfying $\tilde{\mathcal{G}} \subseteq \mathcal{G} \vee \mathcal{B}$, $\mathcal{A} \perp\!\!\!\perp \tilde{\mathcal{G}}|\mathcal{G}$.*

Exercise 7.1.4. *Let $\mathcal{A} \perp\!\!\!\perp \mathcal{B}|\mathcal{G}$. Then*

(i) $\mathcal{A}' \subseteq \mathcal{A} \vee \mathcal{G}$ *and* $\mathcal{B}' \subseteq \mathcal{B} \vee \mathcal{G}$ *imply that* $\mathcal{A}' \perp\!\!\!\perp \mathcal{B}'|\mathcal{G}$,

(ii) $\mathcal{G} \subseteq \mathcal{G}' \subseteq \mathcal{A} \vee \mathcal{B} \vee \mathcal{G}$ *implies that* $\mathcal{A} \perp\!\!\!\perp \mathcal{B}|\mathcal{G}'$.

Let $\{X(\mu), \mu \in M(E)\}$ be a measure indexed random field. Recall that for a subset $S \subseteq E$, we denote its closure, complement, and boundary by $\bar{S}$, S^c, and ∂S, respectively. We define

$$F(S) = \sigma\{X_\mu | \mu \in M(E), \mathrm{supp}(\mu) \subseteq S\},$$

and

$$\Sigma(S) = \bigcap_{S \subseteq \mathcal{O}} F(\mathcal{O})$$

where the intersection is taken over all open sets $\mathcal{O} \subseteq E$ containing S.

Consider an open set D. In [87], McKean defined a Markov property for $\{X_\mu, \mu \in M(E)\}$ as

$$F(D) \perp\!\!\!\perp F(\bar{D}^c)|\Sigma(\partial D). \qquad (7.2)$$

Exercise 7.1.5. *Show that* $F(\overline{D}) \subseteq F(D) \vee \Sigma(\partial D)$ *and* $F(D^c) \subseteq F(\overline{D}^c) \vee \Sigma(\partial D)$. *Use Exercise 7.1.3 to show that McKean condition* (7.2) *implies*

$$F(\overline{D}) \perp\!\!\!\perp F(D^c) | \Sigma(\partial D).$$

We define the Markov property as follows.

Definition 7.1.5. *(Germ Field Markov Property) We say that* $\{X_\mu, \mu \in M(E)\}$ *has Germ Field Markov Property (GFMP) on a subset* $S \subseteq E$ *if*

$$\Sigma(\overline{S}) \perp\!\!\!\perp \Sigma(\overline{S}^c) | \Sigma(\partial S).$$

Exercise 7.1.6. *Use Exercise 7.1.3, part (b) to show that for a stochastic process* $\{X_s, s \in \mathbb{R}\}$, *the Markov property on* $(-\infty, t]$ *implies GFMP on* $S = (-\infty, t]$.

Exercise 7.1.7. *Show that if a stochastic process* $\{X_s, s \in \mathbb{R}\}$ *is continuous, then the McKean Markov property implies GFMP.*

We now derive some results from which we can obtain an alternate form of GMFP.

Lemma 7.1.4. *If* $\{X_\mu, \mu \in M(E)\}$, *is a stochastic random field and* $\mathcal{O}, \mathcal{O}' \subseteq E$ *are open subsets, then* $F(\mathcal{O} \cup \mathcal{O}') = F(\mathcal{O}) \vee F(\mathcal{O}')$.

Proof. Since $F(\mathcal{O})$ and $F(\mathcal{O}')$ are included in $F(\mathcal{O} \cup \mathcal{O}')$, we only need to prove that $F(\mathcal{O} \cup \mathcal{O}') \subseteq F(\mathcal{O}) \vee F(\mathcal{O}')$. Let $\mu \in M(E)$ have its support in $\mathcal{O} \cup \mathcal{O}'$, then by condition (A.2), $\mu = \mu_1 + \mu_2$, where $\mu_1, \mu_2 \in M(E)$, and $\text{supp}(\mu_1) \subseteq \mathcal{O}$, $\text{supp}(\mu_2) \subseteq \mathcal{O}'$. Then $X_\mu = X_{\mu_1} + X_{\mu_2}$, giving the result. $\square$

Lemma 7.1.5. *Let* $\{X_\mu, \mu \in M(E)\}$, *be a stochastic random field and* $S \subseteq E$ *be an open subset. If* $\partial S \subseteq \mathcal{O}$, *then*

$$F(S) \perp\!\!\!\perp F(\overline{S}^c) | F(\mathcal{O}) \text{ if and only if } \Sigma(S) \perp\!\!\!\perp \Sigma(\overline{S}^c) | F(\mathcal{O}).$$

Proof. It is enough to prove the necessity since $F(S) \subseteq \Sigma(\overline{S})$ and $F(\overline{S}^c) \subseteq \Sigma(S^c)$. As $F(S) \vee F(\mathcal{O}) = F(S \cup \mathcal{O}) \supseteq \Sigma(\overline{S})$ and

$$F(\overline{S}^c) \vee F(\mathcal{O}) = F(\overline{S}^c \cup \mathcal{O}) \supseteq \Sigma(S^c),$$

then the result follows by Exercise 7.1.2. $\square$

Exercise 7.1.8. *Use Lemma 7.1.5 and the Martingale Convergence Theorem for net indexed* σ-*fields, [12] to show that*

$$\Sigma(\overline{S}) \perp\!\!\!\perp \Sigma(S^c) | \Sigma(\partial S).$$

Conclude that, under the condition that $\Sigma(\partial S) = \Sigma^-(\partial S) = \Sigma^+(\partial S)$ *the McKean Markov Property implies GFMP [104], where we denote* $\Sigma^-(\partial S) = \bigcap_{\mathcal{O} \supseteq \partial S} F(\mathcal{O} \cap S)$ *and* $\Sigma^+(\partial S) = \bigcap_{\mathcal{O} \supseteq \partial S} F(\mathcal{O} \cap \overline{S})$.

Also, show that if $\{X(\mu),\, \mu \in M(E)\}$ *has GFMP on an open set* $S \subseteq E$, *and if*

$$\Sigma\left(\overline{S}\right) \vee \Sigma\left(S^c\right) = F(E)$$

then

$$\Sigma\left(\overline{S}\right) \perp\!\!\!\perp \Sigma\left(S^c\right) \mid F(\mathcal{O})$$

for all open sets $\mathcal{O} \supseteq \partial S$, *as* $\Sigma(\partial S) \subseteq F(\mathcal{O}) \subseteq \Sigma\left(\overline{S}\right) \vee \Sigma\left(S^c\right)$.

As we shall study Markov properties for Gaussian Random fields, we need to derive the conditions for a Gaussian random field $\{X_\mu,\, \mu \in M(E)\}$ to be Markov. We are going to use conditional probability and projection, and hence we assume that all random fields involved are defined for all sets of measure zero in $\mathcal{F}$. This does affect the definition of conditional independence. We recall that some elementary properties of Gaussian space, conditional expectations, and projections were introduced in Theorem 2.2.2 of Chapter 2. We also include the following property of conditional expectation as an exercise.

Exercise 7.1.9. *Let* H_0, H_1, *and* H_2 *be subspaces of a Gaussian space* H. *Then*

$$\sigma(H_1) \perp\!\!\!\perp \sigma(H_2) \mid \sigma(H_0)$$

if and only if $H_1' \ominus H_0 \perp H_2' \ominus H_0$, *where* H_i' *is a subspace of* H *generated by* $H_i' \oplus H_0$, $i = 1, 2$.

Hint: By Exercise 7.1.4, $\sigma(H_1') \perp\!\!\!\perp \sigma(H_2') \mid \sigma(H_0)$. *Conversely, use the fact that if* $\sigma(H_1' \ominus H_0)$ *is independent of* $\sigma(H_2' \ominus H_0)$, *then* $\sigma(H_1') \perp\!\!\!\perp \sigma(H_2') \mid \sigma(H_0)$.

Lemma 7.1.6. *Suppose* H_0, H_1, H_2 *are subspaces of a Gaussian space, such that* $H_0 \subseteq H_1 \cap H_2$. *Then* $\sigma(H_1) \perp\!\!\!\perp \sigma(H_2) \mid \sigma(H_0)$ *if and only if* $H_1 \cap H_2 = H_0$ *and* $H_1^\perp \perp H_2^\perp$, *where* $H_i^\perp$ *are orthogonal complements of* H_i *in* $H_1 \vee H_2 = \overline{\mathrm{span}\{H_1, H_2\}}$, $(i = 1, 2)$.

Proof. By Exercise 7.1.9, $\sigma(H_1) \perp\!\!\!\perp \sigma(H_2) \mid \sigma(H_0)$ if and only if $H_1 \vee H_2 = (H_1 \ominus H_0) \oplus H_0 \oplus (H_2 \ominus H_0)$. This equality is equivalent to $H_1 \cap H_2 = H_0$ and $H_1^\perp \perp H_2^\perp$. □

As a consequence, we can see that $\sigma(H_1) \perp\!\!\!\perp \sigma(H_2) \mid \sigma(H_1 \cap H_2)$ if and only if $P_{H_1} \circ P_{H_2} = P_{H_1 \cap H_2}$.

Let us now derive the Markov property for Gaussian random fields. We need the following theorem.

Theorem 7.1.1. *Let* $(\Omega, \mathcal{F}, P)$ *be a complete probability space.*

(a) *If* H_1 *and* H_2 *are two subspaces of a Gaussian space* $H \subseteq L^2(\Omega, \mathcal{F}, P)$, *then* $\sigma(H_1 \cap H_2) = \sigma(H_1) \cap \sigma(H_2)$.

(b) *If* $\{H_i,\, i \in I\}$ *are Gaussian subspaces of a Gaussian space* H, *then*

$$\sigma\left(\bigcap_{i \in I} H_i\right) \underset{P}{=} \bigcap_{i \in I} \sigma(H_i).$$

Here, $\underset{P}{=}$ denotes that the two sides are equal except the sets of P measure zero.

Proof. Let $Y \in H$, then

$$E\left(Y \mid \sigma(H_1) \cap \sigma(H_2)\right) = \lim_{n \to \infty} \left(P_{H_1} \circ P_{H_2}\right)^n (Y) \in H_1 \cap H_2,$$

where P_{H_i} is the orthogonal projection on $L_2(\Omega, \sigma(H_i), P)$, $(i = 1, 2)$ because of the Alternating Projection Theorem of Von Neumann [121].

Let g be bounded and $\sigma(H)$-measurable real-valued function, then $g \in L^2(\Omega, \sigma(H), P)$. In view of the homogeneous chaos expansion, Theorem 2.4.1 of Chapter 2, it suffices to prove that

$$E\left(Y_1^{\gamma_1} \cdot \ldots \cdot Y_n^{\gamma_n} \mid \sigma(H_1) \cap \sigma(H_2)\right)$$

is $\sigma(H_1 \cap H_2)$ measurable. Let $Y_i = X_i + Z_i$, where $X_i = E(Y_i \mid \sigma(H_1) \cap \sigma(H_2))$ and Z_i is independent of $\sigma(H_1) \cap \sigma(H_2)$ (see the proof of Proposition 2.2.2). Then

$$Y_1^{\gamma_1} \cdots Y_n^{\gamma_n} = \prod_{i=1}^{n} (X_i + Z_i)^{\gamma_i}$$

are polynomials in $X_1, \ldots, X_n$ with coefficients that are functions of $Z_1, \ldots, Z_n$. Hence, $E\left(Y_1^{\gamma_1} \cdots Y_n^{\gamma_n} \mid \sigma(H_1) \cap \sigma(H_2)\right)$ is in $\sigma(H_1 \cap H_2)$.

To prove part (b), we can assume that $(I, \leq)$ is a directed set. Then for $Y \in H$,

$$E\left(Y \mid \bigcap_{i \in I} \sigma(H_i)\right) = \lim_{i} P_{\bigcap_{j \leq i} H_j} Y \in \bigcap_{i \in I} H_i.$$

Repeating the argument involving the homogeneous chaos expansion, we obtain that

$$E\left(Y_1^{\gamma_1} \cdots Y_n^{\gamma_n} \mid \bigcap_{i \in I} \sigma(H_i)\right) \in \sigma\left(\bigcap_{i \in I} H_i\right).$$

$\square$

We denote for a subset $S \subseteq E$,

$$H(X : S) = \overline{\mathrm{span}}\left\{X_\mu, \mathrm{supp}(\mu) \subseteq S\right\},$$

and

$$\overline{H}(X : S) = \bigcap_{\mathcal{O} \supseteq S} \overline{\mathrm{span}}\left\{X_\mu, \mathrm{supp}(\mu) \subseteq \mathcal{O}\right\}, \; \mathcal{O} \text{ are open.}$$

From Definition 7.1.5, a Gaussian random field has GFMP on $A \subseteq E$ if and only if

$$\bigcap_{\mathcal{O} \supseteq \overline{A}} \sigma\{H(X : \mathcal{O})\} \perp\!\!\!\perp \bigcap_{\mathcal{O} \supseteq \overline{A^c}} \sigma\{H(X : \mathcal{O})\} \Big| \bigcap_{\mathcal{O} \subseteq \partial A} \sigma\{H(X : \mathcal{O})\}.$$

By Theorem 7.1.1, the property GFMP for a Gaussian random field is equivalent to the following condition:

$$P_{\overline{H}(X:\overline{A})} P_{\overline{H}(X:\overline{A^c})} = P_{\overline{H}(X:\partial A)}.$$

7.2 Analytic Conditions for Markov Property of a Measure-Indexed Gaussian Random Field

Let $X = \{X_\mu, \mu \in M(E)\}$ be a Gaussian random field and recall from Section 7.1 that for $S \subseteq E$, we denote $H(X:S) = \overline{\text{span}}\{X_\mu, \text{supp}(\mu) \subseteq S\}$, where the closure is in $L^2(\Omega, \mathcal{F}, P)$. Since $F(S) = \sigma\{H(X:S)\}$, we conclude by Lemmas 7.1.5 and 7.1.6 that X has GFMP on all open set S if and only if

$$H(X:S\cup\mathcal{O})\cap H(X:S^c\cup\mathcal{O}) = H(X:\mathcal{O}) \tag{7.3}$$

and

$$H(X:S\cup\mathcal{O})^\perp \perp H(X:S^c\cup\mathcal{O})^\perp \tag{7.4}$$

for all open sets $S \subseteq E$ and open sets $E \supseteq \mathcal{O} \supseteq \partial S$.

Exercise 7.2.1. *Prove that conditions (7.3) and (7.3) are equivalent to X possessing GFMP property.*

Let us also recall that if C is a covariance of X and $K(C)$ is its RKHS, then we denote by $H(X) = H(E)$ the linear subspace of $L^2(\Omega, \mathcal{F}, P)$ generated by $X = \{X_\mu, \mu \in M(E)\}$, and $\pi: K(C) \to H(E)$ denotes the linear isometry defined by $\pi(C(\cdot, \mu)) = X_\mu$. Hence, if $S \subseteq E$, we denote

$$K(S) = \pi^{-1}(H(X:S)) = \overline{\text{span}}\{C(\cdot, \mu) \mid \text{supp}(\mu) \subseteq S\},$$

where the closure is taken in $K(C)$. By Definition 7.1.1, the support $\text{supp}(f)$ of a linear functional $f: M(E) \to \mathbb{R}$ is defined as the largest open set $\mathcal{O} \subseteq E$, such that $f(\mu) = 0$ for all $\mu \in M(E)$ with $\text{supp}(\mu) \subseteq \mathcal{O}$.

Our first result relates GMFP to the properties of the elements in $K(C)$.

Theorem 7.2.1. *Let E be a locally compact Hausdorff second-countable topological space and $M(E)$ be a space of Radon measures on E with compact support satisfying conditions (A1) and (A2) of Section 7.1. Then the centered Gaussian process $\{X_\mu, \mu \in M(E)\}$ has GFMP on all open sets if and only if the following conditions are satisfied:*

(a) *For any $f_1, f_2 \in K(C)$ with $\text{supp}(f_1) \cap \text{supp}(f_2) = \varnothing$, $\langle f_1, f_2 \rangle_{K(C)} = 0$.*

(b) *If $f \in K(C)$ and $f = f_1 + f_2$, where f_1 and f_2 as linear functionals on $M(E)$ have disjoint supports, then $f_1, f_2 \in K(C)$.*

Proof. Let us suppose that conditions (a) and (b) hold. To verify (7.3), it is enough to prove that for any open set $\mathcal{O} \supseteq \partial S$,

$$H(X:S\cup\mathcal{O})\cap H\left(X:\overline{S}^c\cup\mathcal{O}\right) \subseteq H(X:\mathcal{O}), \tag{7.5}$$

or equivalently that

$$K(\mathcal{O})^\perp \subseteq K(S\cup\mathcal{O})^\perp \vee K(\overline{S}^c\cup\mathcal{O})^\perp, \tag{7.6}$$

(subspace of $K(E)$ generated by $K(S \cup \mathcal{O})^{\perp}$ and $K(\overline{S}^c \cup \mathcal{O})^{\perp}$). Let $f \in K(\mathcal{O})^{\perp}$. Then $f(\mu) = \langle f, C(\cdot, \mu) \rangle_{K(C)} = 0$ if $\mathrm{supp}(\mu) \subseteq \mathcal{O}$, hence $\mathrm{supp}(f) \subseteq \mathcal{O}^c$. Observe that

$$\overline{S} \cap \mathcal{O}^c = (S \cap \partial S) \cap \mathcal{O}^c = S \cap \mathcal{O}^c$$

and

$$\overline{S}^c \cap \mathcal{O}^c = \overline{\overline{S}^c} \cap \mathcal{O}^c \quad (\partial \overline{S}^c \subseteq \partial S \subseteq \partial \overline{S} \subseteq \mathcal{O}),$$

showing that $S \cap \mathcal{O}^c$ and $\overline{S}^c \cap \mathcal{O}^c$ are disjoint closed sets whose union is $\mathcal{O}^c$. By (A2),

$$f(\mu) = f_1(\mu) + f_2(\mu)$$

with f_1 and f_2 being linear functionals on $M(E)$ and $\mathrm{supp}(f_1) \subseteq S \cap \mathcal{O}^c$, $\mathrm{supp}(f_2) \subseteq \overline{S}^c \cap \mathcal{O}^c$.

By condition (b), $f_1, f_2 \in K(C)$. Then

$$f_1 \in K\left(\left(\overline{S} \cap \mathcal{O}^c\right)^c\right)^{\perp} = K\left(\overline{S}^c \cup \mathcal{O}\right)^{\perp}$$

and

$$f_2 \in K\left(\left(\overline{S}^c \cap \mathcal{O}^c\right)^c\right)^{\perp} = K\left(\overline{S} \cup \mathcal{O}\right)^{\perp} = K(S \cup \mathcal{O}) \perp.$$

To show (7.6) and conclude (7.5), it is enough to prove that

$$K(S \cup \mathcal{O})^{\perp} \perp K(\overline{S}^c \cup \mathcal{O})^{\perp}. \tag{7.7}$$

This follows, if we show that

$$K(S \cup \mathcal{O})^{\perp} \subseteq \overline{\mathrm{span}}\left\{ f \in K(C) \,\middle|\, \mathrm{supp}(f) \subseteq \overline{S}^c \right\}$$

and

$$K(\overline{S}^c \cup \mathcal{O})^{\perp} \subseteq \overline{\mathrm{span}}\left\{ f \in K(C) \mid \mathrm{supp}(f) \subseteq S \right\}.$$

We observe that if $f \in K(S \cup \mathcal{O})^{\perp}$, then for all $\mu \in M(E)$ with $\mathrm{supp}(\mu) \subseteq S \cup \mathcal{O}$, $f(\mu) = 0$. Hence,

$$\mathrm{supp}(f) \subseteq (S \cup \mathcal{O})^c = \left(\overline{S} \cup \mathcal{O}\right)^c = \overline{S}^c \cap \mathcal{O}^c \subseteq \overline{S}^c.$$

Similarly , if $f \in K(\overline{S}^c \cup \mathcal{O})^{\perp}$ then

$$\mathrm{supp}(f) \subseteq \left(\overline{S}^c \cup \mathcal{O}\right)^c = \overline{S} \cap \mathcal{O}^c \subseteq S.$$

To prove the necessity of conditions (a) and (b), assume that (7.3) and (7.4) hold. Let $f_1, f_2 \in K(C)$ have disjoint supports. Then there exists an open set S such that $\mathrm{supp}(f_1) \subseteq S$ and $\mathrm{supp}(f_2) \subseteq \overline{S}^c$ and $\mathcal{O} = (\mathrm{supp}(f_1) \cup \mathrm{supp}(f_2))^c$ is an open set containing ∂S. Since $S \cup \mathcal{O} \subseteq (\mathrm{supp}(f_2))^c$ and $\overline{S}^c \cup \mathcal{O} \subseteq (\mathrm{supp}(f_1))^c$, we conclude that $f_1 \in K\left(\overline{S}^c \cup \mathcal{O}\right)^{\perp}$ and $f_2 \in K(S \cup \mathcal{O})^{\perp}$, giving $\langle f_1, f_2 \rangle_{K(C)} = 0$

by (7.4), which proves (a) is necessary. To prove that (b) must hold true, assume $f \in K(C)$ and $f = f_1 + f_2$, where f_1 and f_2 are linear functionals on $M(E)$ having disjoint supports. We choose an open set S, such that $\text{supp}(f_1) \subseteq S$ and $\text{supp}(f_2) \subseteq \overline{S}^c$. Let $\mathcal{O} = (\text{supp}(f_1) \cap \text{supp}(f_2))^c$. Then by (7.3) and (7.4) we have

$$H(X) = H(X : S \cap \mathcal{O})^{\perp} \oplus H\left(X : \overline{S}^c \cup \mathcal{O}\right)^{\perp} \oplus H(X : \mathcal{O}),$$

or equivalently

$$K(C) = K(S \cap \mathcal{O})^{\perp} \oplus K\left(\overline{S}^c \cup \mathcal{O}\right)^{\perp} \oplus K(\mathcal{O}).$$

Note that $\text{supp}(f) \subseteq \text{supp}(f_1) \cup \text{supp}(f_2) \subseteq \mathcal{O}^c$, giving $f \in K(\mathcal{O})^{\perp}$. Hence $f = f'_1 + f'_2$, where $f'_1 \in K\left(\overline{S}^c \cup \mathcal{O}\right)^{\perp}$ and $f'_2 \in K(S \cup \mathcal{O})^{\perp}$. This implies that

$$\begin{aligned}
\text{supp}(f'_1) \subseteq \left(\overline{S}^c \cup \mathcal{O}\right)^c &= \overline{S} \cap \mathcal{O}^c \\
&= \overline{S} \cap (\text{supp}(f_1) \cup \text{supp}(f_2)) \\
&= (\overline{S} \cap \text{supp}(f_1)) \cup (\overline{S} \cap \text{supp}(f_2)) \\
&= \overline{S} \cap \text{supp}(f_1) \subseteq \text{supp}(f_1).
\end{aligned}$$

Similarly, $\text{supp}(f'_2) \subseteq \text{supp}(f_2)$. Now $f = f_1 + f_2 = f'_1 + f'_2, f_1 - f'_1 = f'_2 - f_2$ with $\text{supp}(f_1 - f'_1) \cap \text{supp}(f'_2 - f_2) = \varnothing$, which implies $f_1 = f'_1$ and $f_2 = f'_2$ by virtue of part (b) of Lemma 7.1.1. This gives $f_1, f_2 \in K(C)$. □

If we only consider the GFMP on all open subsets that are bounded or have bounded complements, then we have the following theorem similar to Theorem 7.2.1.

Theorem 7.2.2. *Let E and $M(E)$ be as in Theorem 7.2.1. Then the centered Gaussian random field $\{X_{\mu}, \mu \in M(E)\}$ has GFMP on all open subsets of E that are bounded or have bounded complements if and only if the following conditions hold true:*

(a) *If $f_1, f_2 \in K(C)$ with $\text{supp}(f_1) \cap \text{supp}(f_2) = \varnothing$, and at least one of the sets $\text{supp}(f_i)$, $(i = 1, 2)$ is compact, then $\langle f_1, f_2 \rangle_{K(C)} = 0$. (This condition is called compact local property of the RKHS $K(C)$)*

(b) *If $f \in K(C)$ and $f = f_1 + f_2$, where both f_1, f_2 are linear functionals on $M(E)$ with $\text{supp}(f_1) \cap \text{supp}(f_2) = \varnothing$, and at least one of $\text{supp}(f_i)$, $(i = 1, 2)$ is compact, then $f_1, f_2 \in K(C)$.*

Exercise 7.2.2. *Prove Theorem 7.2.2. Hint: The arguments in the proof of Theorem 7.2.1 go through. Note that if $\text{supp}(f_1)$ is compact, then one can choose a pre-compact set S to cover $\text{supp}(f_1)$ and $\overline{S}^c$ covers $\text{supp}(f_2)$.*

We extend the concept of the dual process introduced by [88] (see [81]) to the measure-indexed random field $\{X_{\mu}, \mu \in M(E)\}$ defined on a probability space $(\Omega, \mathcal{F}, P)$. Let $G(E)$ be a subset of $C_0(E)$ ($G(E)$ does not need to be a linear subspace) and $\hat{X} = \{\hat{X}_g, \ g \in G(E)\}$ be a Gaussian random field defined on the same probability space.

Definition 7.2.1. *The Gaussian random field* $\hat{X} = \{\hat{X}_g, \ g \in G(E)\}$ *is called a dual (or biorthogonal) field of* $X = \{X_\mu, \ \mu \in M(E)\}$ *if* $H(X) = H(\hat{X})$, *and*

$$E\hat{X}_g X_\mu = \int_E g \, d\mu$$

for all $g \in G(E)$ *and* $\mu \in M(E)$, *where* $H(X)$ *and* $H(\hat{X})$ *are the linear subspaces of* $L_2(\Omega, \mathcal{F}, P)$ *generated by* $\{X_\mu, \ \mu \in M(E)\}$ *and* $\{\hat{X}_g, \ g \in G(E)\}$, *respectively.*

For any $g \in G(E)$ let

$$f_g(\mu) = \int_E g \, d\mu, \quad \mu \in M(E).$$

Assuming that the dual field $\hat{X}$ to X exists, clearly $f_g \in K(C)$. For any open subset $D \subseteq E$, we define the following subspaces of $K(C)$:

$$
\begin{aligned}
M(D) &= \overline{\text{span}}\{f \mid f \in K(C) \text{ and } \text{supp}(f) \subseteq D\} \\
\hat{M}(D) &= \overline{\text{span}}\{f_g \mid g \in G(E) \text{ and } \text{supp}(g) \subseteq D\},
\end{aligned}
$$

where $\text{supp}(g)$ is the closure of the set $\{e, \ g(e) \neq 0\}$ in E. We observe that

$$\hat{M}(D) \subseteq M(D) \subseteq K(D^c)^\perp$$

for every open subset $D \subseteq E$.

Definition 7.2.2. *We say that* $G(E)$ *has partition of the unity property if for every* $g \in G(E)$ *and an open covering* $\{\mathcal{O}_1, ..., \mathcal{O}_n\}$ *of* $\text{supp}(g)$, *we have a representation* $g = \sum_i^n = g_i$, *with* $g_i \in G(E)$, *and* $\text{supp}(g_i) \subseteq \mathcal{O}_i$, $(i = 1, ..., n)$.

Lemma 7.2.1. *If* $G(E)$ *has the partition of unity property, then* $\hat{M}(D_1 \cup D_2) = \hat{M}(D_1) \vee \hat{M}(D_2)$ *for any open subsets* $D_1, D_2 \subseteq E$.

Proof. Let $g \in G(E)$ with $\text{supp}(g) \subseteq D_1 \cup D_2$. Then $g = g_1 + g_2$, with $g_i \in G(E)$, and $\text{supp}(g_i) \subseteq D_i$, $(i = 1, 2)$. Hence

$$f_g(\mu) = f_{g_1}(\mu) + f_{g_2}(\mu), \quad \text{for } \mu \in M(E)$$

with $f_{g_i} \in \hat{M}(D_i)$, $(i = 1, 2)$, giving that $f_g \in \hat{M}(D_1) \vee \hat{M}(D_2)$, and the result follows. $\square$

Theorem 7.2.3. *Let* $\hat{X}$ *be a dual field of* X, *such that* $G(E)$ *has the partition of unity property and* $\hat{M}(D) = M(D)$ *for all open subsets* $D \subseteq E$. *Then, condition (a) of Theorem 7.2.1 implies condition (b) of that theorem.*

Proof. By condition (a) of Theorem 7.2.1, $M(D_1) \perp M(D_2)$ for every two disjoint open subsets $D_1, D_2 \subseteq E$, and by Lemma 7.2.1 and the assumption $\hat{M}(D) = M(D)$, we conclude that

$$M(D_1 \cup D_2) = \hat{M}(D_1 \cup D_2) = \hat{M}(D_1) \vee \hat{M}(D_2) = M(D_1) \oplus M(D_2).$$

Let $f \in K(C)$, with $f = f_1 + f_2$, where f_1, f_2 are linear functionals on $M(E)$ with $\text{supp}(f_1) \cap \text{supp}(f_2) = \varnothing$. Then we can choose open sets D_1, D_2, such that $\text{supp}(f_i) \subseteq D_i$, $(i = 1, 2)$ and $\overline{D}_1, \overline{D}_2$ are disjoint. Let $D = D_1 \cup D_2$. Then $\text{supp}(f) \subseteq D$, hence $f \in M(D)$, so that we can write

$$f = P_{M(D)}f = P_{M(D_1)}f + P_{M(D_2)}f = f_1' + f_2'$$

$P_{M(D)}, P_{M(D_i)}$, $(i = 1, 2)$ denote orthogonal projections, and $f_i' \in M(D_i)$, $(i = 1, 2)$. Consider f_1', as an element of $M(D_1)$ it is the limit of a sequence $\{\tilde{f}_n\} \subseteq K(C)$, such that $\text{supp}(\tilde{f}_n) \subseteq D_1$. Hence, for any $\mu \in M(E)$ with $\text{supp}(\mu) \subseteq \overline{D}_1^c$ $f_1'(\mu) = \lim_n \tilde{f}_n(\mu) = 0$, giving that $\text{supp}(f_1') \subseteq \overline{D}_1$. Similarly, $\text{supp}(f_2') \subseteq \overline{D}_2$.

Now $f = f_1 + f_2 = f_1' + f_2'$ gives $f_1 - f_1' = f_2' - f_2$. But $\text{supp}(f_i - f_i') \subseteq \overline{D}_i$, $(i = 1, 2)$ and $\overline{D}_1, \overline{D}_2$ are disjoint. This implies $\text{supp}(f_1 - f_1') = \text{supp}(f_2' - f_2) = \varnothing$. Hence, $f_i = f_i' \in K(C)$, $(i = 1, 2)$. $\square$

The following theorem can be proved in a similar way as Theorem 7.2.3.

Theorem 7.2.4. *Let $\hat{X}$ be dual field of X such that $G(E)$ has the partition of unity property and $\hat{M}(D) = M(D)$ for all open sets that are bounded or have bounded complements. Then, condition (a) of Theorem 7.2.2 implies condition (b) of that theorem.*

Exercise 7.2.3. *Prove Theorem 7.2.4.*

The next corollary is an immediate consequence of Theorem 7.2.1 and Theorem 7.2.2.

Corollary 7.2.1. *Under the assumptions of Theorem 7.2.3 X has GFMP on all open sets if and only if $K(C)$ has the compact local property.*

We will now discuss a specific case when a specific condition on $M(E)$, $G(E)$, and the norm in $K(C)$ can be given in order that $M(D) = \hat{M}(D)$ for the class of all open sets or the class of all open sets that are bounded or have bounded complements ([111], p. 108).

Example 7.2.1 (Duality). *Assume as in Example 7.1.1 that $E \subseteq \mathbb{R}^n$ is an open subset, and*

$$M(E) = \{\mu \mid d\mu = \varphi dx, \ \varphi \in C_0^\infty(E)\}.$$

Let C be covariance on $M(E)$ and denote by $H(E) = H(C)$ the corresponding Gaussian space and by $K(E) = K(C)$ the RKHS of C.

For a subset $S \subseteq E$, $H(S)$ denotes the Gaussian space generated by $C(\cdot, \varphi dx)$ with $\text{supp}(\varphi) \subseteq S$. Define for an ε-neighborhood of $S \subseteq E$,

$$H_+(S) = \bigcap_{\varepsilon > 0} H(S^\varepsilon). \tag{7.8}$$

In [111] Rozanov defines a dual random field $\hat{H}(S)$ for a complete system $\mathcal{G}$ of open sets $S \subseteq E$. A system $\mathcal{G}$ of open sets $S \subseteq E$ is complete if

(i)   *$\mathcal{G}$ contains all open sets which are relatively compact or have relatively compact complement,*

(ii) *$\mathcal{G}$ contains all sufficiently small ε-neighborhoods of the boundary ∂S for any $S \in \mathcal{G}$,*

(iii) *if $S' \in \mathcal{G}$ and $S'' \in \mathcal{G}$, then $S' \cup S'' \in \mathcal{G}$,*

(iv) *if $S \in \mathcal{G}$, then $E \setminus \overline{S} \in \mathcal{G}$.*

For $S \in \mathcal{G}$, $S \subseteq E$, Rozanov defines a random field $\hat{H}(S)$, $S \subseteq E$ to be dual to the random field $H(S)$, $S \subseteq E$ on the complete system $\mathcal{G}$ if

$$\hat{H}(E) \;=\; H(E)$$

and   (7.9)

$$\hat{H}(S) \;=\; H_+(S^c)^\perp, \quad \text{for every} \quad S \in \mathcal{G}.$$

We add the following condition to the definition of duality. Two random fields X_μ and $\hat{X}_\nu$ are biorthogonal if $H(E) = \hat{H}(E)$ and

$$E\left(X_\mu \hat{X}_\nu\right) = \int_E f(x)g(x)\,dx, = \int_E g(x)\mu(dx),$$ (7.10)

where $\mu = f(x)\,dx$, and $\nu = g(x)\,dx$.

If we consider biorthogonal random fields X_μ and $\hat{X}_\nu$, then (7.10) implies that

$$\hat{H}(S) \subseteq H_+(S^c)^\perp = \left[\bigcap_{\varepsilon > 0} H(S^c)^\varepsilon\right]^\perp.$$ (7.11)

since if $g \in C_0^\infty(S)$, g has a compact support, which must be a positive distance from S^c, S being open. Therefore, $\operatorname{supp}(g)$ must be outside of some $(S^c)^\varepsilon$. Hence

$$E\left(X_{f(x)\,dx}\hat{X}_{g(x)\,dx}\right) = \int_E f(x)g(x)\,dx = 0$$

for all $f \in C_0^\infty\left((S^c)^\varepsilon\right)$ and thus $X_{f(x)\,dx} \perp \hat{X}_{g(x)\,dx}$ for all $f \in C_0^\infty\left((S^c)^\varepsilon\right)$, giving $\hat{X}_g(x)\,dx \perp H\left[(S^c)^\varepsilon\right]$, so that $\hat{X}_g(x)\,dx \perp H_+(S)$. Denote by $S^{-\varepsilon}$ the set of points in S whose distance from ∂S (boundary of S) is greater than ε,

$$S^{-\varepsilon} = E \setminus \overline{(S^c)^\varepsilon}.$$

Because of the isomorphism (7.10), the set

$$\left[H\left((S^c)^\varepsilon\right)\right]^\perp = \left\{\hat{X}_{g(x)\,dx} \,\Big|\, \operatorname{supp}(g) \subseteq \overline{(S^{-\varepsilon})}\right\},$$ (7.12)

is unitarily isomorphic to the space of elements $\nu \in K(E)$ (the RKHS of X) such

that $\mathrm{supp}(v) \subseteq \overline{(S^{-\varepsilon})}$ *(meaning* $v(g) = 0$ *for all g with the support in* $\overline{(S^{-\varepsilon})^c}$*).*
Hence,

$$[H_+(S^c)]^\perp = \bigvee_{\varepsilon>0} \left\{ \hat{X}_{g(x)\,dx} \,\middle|\, \mathrm{supp}(g) \subseteq \overline{(S^{-\varepsilon})} \right\} \tag{7.13}$$

is unitarily isomorphic to the closure of all $v \in K(E)$ *with the support in* $\overline{(S^{-\varepsilon})}$,
$\varepsilon > 0$, *because*

$$[H_+(S^c)]^\perp = \bigvee_{\varepsilon>0} H\left[(S^c)^\varepsilon\right]^\perp. \tag{7.14}$$

Denote by $V(S)$ *the closure in* $K(E)$ *of the elements* $v \in K(E)$, *with* $\mathrm{supp}(v) \subseteq$
$\overline{(S^{-\varepsilon})}$, $\varepsilon > 0$. *Observe that biorthogonality implies that, in the sense of identification of* $f(x)$ *with* $C(\cdot, f(x)\,dx)$

$$C_0^\infty(E) \subseteq K(E), \quad \overline{C_0^\infty(E)} = K(E),$$

so that every element $v \in K(E)$ *can be identified with a limit of some sequence*
f_n *of elements of* $C_0^\infty(E)$. *If we also could claim that*

$$\overline{C_0^\infty(S)} = V(S), \tag{7.15}$$

then using the fact that in the sense of unitary isomorphism

$$\overline{C_0^\infty(S)} = \hat{H}(S)$$

we can conclude that

$$\hat{H}(S) = V(S) = H_+(S^c)^\perp$$

thus obtaining the duality condition (7.9). The following lemma (pg. 108 in [111]) provides a condition for (7.15) to hold true.

Lemma 7.2.2. *Suppose that the operation of multiplication by a function* $w \in C_0^\infty(\mathbb{R}^d)$ *is bounded in the space* $K(E)$, *that is*

$$\|wf\|_{K(E)} \le M \|f\|_{K(E)}, \quad f \in C_0^\infty(E) \tag{7.16}$$

where

$$\|f\|_{K(E)} = \|C(\cdot, f(x)\,dx))\|_K(E).$$

Then condition (7.15) is satisfied for all subsets $S \subseteq E$ *that are bounded or have bounded complement in E.*

Proof. We identify where we identify elements of $\overline{C_0^\infty(E)}$ with elements of $K(E)$. Since $\overline{C_0^\infty(E)} = K(E)$, then for any $v \in K(E)$ with $\mathrm{supp}(v) \subseteq \overline{(S^{-\varepsilon})}$, there exists a sequence of functions $f_n \in C_0^\infty(E)$, such that $f_n \to v$ in $K(E)$. For a bounded set $S \subseteq E$, there exists $w \in C_0^\infty(\mathbb{R}^d)$, such that for $0 < \delta < \varepsilon$,

$$w(t) = \begin{cases} 1 & \text{if } t \in \overline{(S^{-\varepsilon})} \\ 0 & \text{if } t \in (S^c)^\delta \end{cases}$$

Then $v_n = w f_n \in C_0^\infty(S)$ and by (7.16),

$$\|v_n\|_{K(E)} \leq M \|f_n\|_{K(E)} \leq M \sup_n \|f_n\|_{K(E)} < \infty.$$

For any $g \in C_0^\infty(E)$, using (7.10),

$$\langle g, v_n \rangle_{K(E)} = \langle w g, f_n \rangle_{K(E)} \to \langle w g, v \rangle_{K(E)} = \langle f, v \rangle_{K(E)},$$

since $\mathrm{supp}(v) \subseteq \overline{(S^{-\varepsilon})}$. Thus, the sequence v_n, being bounded in $K(E)$, converges weakly to $v \in K(E)$. By Mazur's theorem, v can be approximated strongly by convex linear combinations of $v_n \in C_0^\infty(S)$. $\qquad\square$

Exercise 7.2.4. *Prove Lemma 7.2.2 for the case of subsets with bounded complements in E.*

Remark 7.2.1. *We note that Rozanov's condition for duality (7.9) corresponds in our notation to the condition $\hat{M}(D) = M(D)$ for all open sets D bounded or with bounded complements.*

    Using Lemma 7.2.2 and Corollary 7.2.1, we obtain the next result.

Theorem 7.2.5. *Let X be centered Gaussian random field with the dual (and biorthogonal) field $\hat{X}$. If the assumptions of Lemma 7.2.2 are satisfied, then the following are equivalent:*

(a) *The RKHS $K(C)$ has local (respectively, compact local) property.*

(b) *$\hat{M}(D_1) \perp \hat{M}(D_2)$ for all disjoint open (respectively, bounded open) subsets $D_1, D_2 \subseteq E$.*

    This ends our discussion on the specific case discussed in [111].

7.3   Markov Property of Measure-Indexed Gaussian Random Fields Associated with Dirichlet Forms

In this section, we study an interesting example of Gaussian Markov random field. In his pioneering work, Dynkin [21] showed how to associate a Green function of a symmetric multivariate Markov process with a Gaussian Markov random field for which the splitting field can be explicitly described in terms of the multivariate Markov process. Subsequently, Röckner [108] generalized results of Dynkin by considering a symmetric Markov process as described in the work of Fukushima [30] using Dirichlet forms.

    We shall show that the work of Röckner can be derived as a consequence of the general Theorem 7.3.2. In the next section, we present the basic ideas needed for describing the concepts in [108]. The analytic results from [30] used here are collected in Appendices 7.4 and 7.5. We end the chapter by giving concrete examples of interest.

7.3.1   Gaussian Processes Related to Dirichlet Forms

We start with basic concepts and preliminary facts. Let $(H, \langle \cdot, \cdot \rangle_H)$ be a real Hilbert space.

Definition 7.3.1. *A function $\mathcal{E} : H \times H \to \mathbb{R}$ is called a non-negative definite symmetric bilinear form $\mathcal{E}$ densely defined on H if*

(E1) *$\mathcal{E}$ is defined on $\mathcal{D} \times \mathcal{D}$ with values in $\mathbb{R}$ and $\mathcal{D}$ is a dense subspace of H.*

(E2) *$\mathcal{E}$ has the following properties. For any $u, v, w \in \mathcal{D}$ and $a \in \mathbb{R}$,*

$$
\begin{aligned}
\mathcal{E}(u,v) &= \mathcal{E}(v,u) \quad \text{(symmetry)} \\
\mathcal{E}(u+v,w) &= \mathcal{E}(u,w) + \mathcal{E}(v,w) \quad \text{(bilinearity)} \\
a\mathcal{E}(u,v) &= \mathcal{E}(av,u) \\
\mathcal{E}(u,u) &\geq 0 \quad \text{non-negative definiteness}
\end{aligned}
$$

We call $\mathcal{D}$ the domain of $\mathcal{E}$.

We say that the form $\mathcal{E}$ is closed if for any sequence $\{u_n\}_{n=0}^{\infty} \subseteq \mathcal{D}$, if $\mathcal{E}(u_n, u_n)$ converges and $\langle u_n, u_n \rangle_H \to 0$ implies that $\mathcal{E}(u_n, u_n) \to 0$.

We shall consider henceforth $H = L^2(E, m)$, where $\text{supp}(m) = E$ and m is a Radon measure on a locally compact separable Hausdorff space $(E, \mathcal{B}(E))$ (the Borel measure space). We assume that m is finite on compact subsets of E and strictly positive on non-empty open subsets of E. An operator T on a Hilbert space H is *non-positive definite* if for all $h \in H$, $\langle Th, h \rangle_H \leq 0$. It is *called self-adjoint* if for any $h, g \in H$, $\langle Th, g \rangle_H = \langle h, Tg \rangle_H$. The following theorem is in [30].

Theorem 7.3.1. *There is a one-to-one correspondence between the family of closed symmetric bilinear forms $\mathcal{E}$ on H and the family of non-positive self-adjoint linear operators A on H, with*

$$
\mathcal{D}(\mathcal{E}) = \mathcal{D}(\sqrt{-A}) \quad \text{and} \quad \mathcal{E}(u,v) = \langle \sqrt{-A}u, \sqrt{-A}v \rangle_H, \quad u, v \in \mathcal{D}(\mathcal{E}).
$$

This correspondence can be characterized by

$$
\mathcal{D}(\mathcal{E}) \supseteq \mathcal{D}(A) \quad \text{and} \quad \mathcal{E}(u,v) = \langle -Au, v \rangle_H, \quad u, v \in \mathcal{D}(\mathcal{E}).
$$

Definition 7.3.2. *A bounded operator S on $L^2(E, m)$ is called Markovian if $0 \leq Su \leq 1$, m-a.e. whenever $u \in L^2(E, m)$ satisfies $0 \leq u \leq 1$, m-a.e.*

We observe that by Theorem 1.4.1 in [30], if $\{T_t\}_{t \geq 0}$ is a strongly continuous semigroup on $L^2(E, m)$ with the infinitesimal operator A and the corresponding bilinear form $\mathcal{E}$, then

$$
\mathcal{D}(\mathcal{E}) = \left\{ u \in L^2(E,m) \,\middle|\, \lim_{t \to 0} \frac{1}{t} \langle u - T_t u, u \rangle_{L^2(E,m)} < \infty \right\}
$$

and

$$
\mathcal{E}(u,v) = \lim_{t \to 0} \frac{1}{t} \langle u - T_t u, v \rangle_{L^2(E,m)}, \quad u, v \in \mathcal{D}(\mathcal{E}).
$$

Definition 7.3.3. *We say that a strongly continuous semigroup T_t on $L^2(E,m)$ is Markovian if T_t is a Markovian operator for each $t > 0$. If T_t on $L^2(E,m)$ is Markovian, then we call the corresponding Dirichlet form $\mathcal{E}$ Markovian.*

Denote $\mathcal{F} = \mathcal{D}(\mathcal{E})$. If $\mathcal{E}$ is a closed Markovian symmetric form on $L^2(E,m)$, then following [30] (p. 32), we call $(\mathcal{F},\mathcal{E})$ a Dirichlet space relative to $L^2(E,m)$.

If T_t is a transient Markov semigroup (see Appendix 7.4), that is, the associated Markov process is transient, then we introduce an extended (transient) Dirichlet space $(\mathcal{F}_e,\mathcal{E})$ with reference measure m. The reason is that the Dirichlet space $(\mathcal{F},\mathcal{E})$ in general may not even be a pre-Hilbert space, but it can be enlarged (by an abstract completion in $\mathcal{E}$) to an extended space if (and only if) the associated Markov process is transient. Following [108], we shall consider the Gaussian random field associated with a transient Dirichlet space. We begin with the necessary terminology needed for our purpose.

Definition 7.3.4. *A pair $(\mathcal{F}_e,\mathcal{E})$ is called a regular extended (transient) Dirichlet space with reference measure m if the following conditions are satisfied.*

($\mathcal{F}_e$-1) $\mathcal{F}_e$ is a Hilbert space with inner product $\mathcal{E}$.

($\mathcal{F}_e$-2) There exists an m-integrable bounded function g, strictly positive m-a.e. such that $\mathcal{F}_e \subseteq L^1(E, v_g)$ $(v_g = g\,dm)$ and

$$\int_E |u(x)| \, v_g(dx) = \int_E |u(x)| g(x) \, m(dx) \le \sqrt{\mathcal{E}(u,u)}, \quad u \in \mathcal{F}_e.$$

($\mathcal{F}_e$-3) $\mathcal{F}_e \cap C_0(E)$ is dense both in $(\mathcal{F}_e,\mathcal{E})$ and $(C_0(E), \|\cdot\|_{C_0(E)})$, where $C_0(E)$ denotes the space of real-valued continuous functions on E with the supremum norm, $\|f\|_{C_0(E)} = \sup_{x \in E} |f(x)|$.

We say that a function v is a normal contraction of a function u if

$$|v(x) - v(y)| \le |u(x) - u(y)| \quad and \quad |v(x)| \le |u(x)|, \quad x,y \in E.$$

($\mathcal{F}_e$-4) Every normal contraction operates on $(\mathcal{F}_e,\mathcal{E})$, that is, if $u \in \mathcal{F}_e$ and v is a normal contraction of u, then $v \in \mathcal{F}_e$ and $\mathcal{E}(v,v) \le \mathcal{E}(u,u)$.

The function g in condition ($\mathcal{F}_e$-2) is called a reference function of the extended (transient) Dirichlet space $(\mathcal{F}_e,\mathcal{E})$.

We now define the space of measures of bounded energy.

Definition 7.3.5. *A signed measure μ on E is called a measure of bounded energy if there exists a constant $c > 0$, such that*

$$\int_E |u| \, d|\mu| \le c\sqrt{\mathcal{E}(u,u)}$$

for all $u \in \mathcal{F}_e \cap C_0(E)$, here $|\mu| = \max(\mu, -\mu)$.

Denote by

$$M_\mathcal{E} \quad \text{all measures on } E \text{ that are of bounded energy}$$
$$M(E) \ = \{\mu \in M_\mathcal{E} \mid \text{supp}(\mu) \text{ is compact}\}$$
$$M_\mathcal{E}^+ \quad = \{\mu \in M_\mathcal{E} \mid \mu \geq 0\}$$
$$M^+(E) = M_\mathcal{E}^+ \cap M(E)$$

Notice that $M(E)$ is a vector space and $\mu \in M(E)$ implies $1_A \mu \in M(E)$ for all Borel sets $A \subseteq E$. Using Example 7.1.2 we conclude that $M(E)$ satisfies conditions (A1), (A2), and (A3) of Section 7.1.

Remark 7.3.1. *In the sequel we will use 0-order version of the results from [30] ($\alpha = 0$ in the notation in [30], see remark there on pg. 81).*

By Theorem 3.2.2. (pg. 71 in [30]), for every measure $\mu \in M_\mathcal{E}^+$, there exists a unique element $U\mu \in \mathcal{F}_e$, such that

$$\mathcal{E}(U\mu, v) = \int_E \tilde{v} d\mu, \quad \text{for all } v \in \mathcal{F}_e,$$

where $\tilde{v}$ denotes any quasi-continuous modification (see Appendix 7.4) in the restricted sense. Define the map $U : M_\mathcal{E} \to \mathcal{F}_e$ by

$$U\mu = U\mu^+ - U\mu^-,$$

where μ^+ and μ^- are the positive and negative parts of μ in the Jordan decomposition. We call $U\mu$ the *potential* of μ. The map U is linear and its kernel is $\{0\}$. By defining a scalar product and norm on $M_\mathcal{E}$ as follows:

$$\langle \mu, v \rangle_{M_\mathcal{E}} \ = \ \mathcal{E}(U\mu, Uv),$$
$$\|\mu\|_{M_\mathcal{E}} \ = \ \mathcal{E}(U\mu, U\mu),$$

with $\mu, v \in M_\mathcal{E}$, and extending them to the completion of $(M_\mathcal{E}, \langle \cdot, \cdot \rangle_{M_\mathcal{E}})$ we obtain a Hilbert space which is unitary isomorphic to $(\mathcal{F}_e, \mathcal{E})$, since by Theorem 3.2.4 in [30] and condition $(\mathcal{F}_e$–3$)$, potentials are dense in $\mathcal{F}_e$. Hence, we can write

$$\mathcal{F}_e = \overline{\text{span}}^\mathcal{E} \left\{ U\left(M_\mathcal{E}^+\right) - U\left(M_\mathcal{E}^-\right) \right\}. \tag{7.17}$$

Define for $A \subseteq E$,

$$M_E(A) \ = \ \{\mu \in M(E) \mid \text{supp}(\mu) \subseteq A\}$$
$$M_\mathcal{E}(A) \ = \ \{\mu \in M_\mathcal{E} \mid \text{supp}(\mu) \subseteq A\}.$$

We have the following lemma.

Lemma 7.3.1. *Let $A \subseteq E$, then*

$$\overline{\text{span}}^{\mathcal{E}} \{ U\mu \mid \mu \in M_{\mathcal{E}}(A) \} = \overline{\text{span}}^{\mathcal{E}} \{ U\mu \mid \mu \in M_E(A) \}.$$

In particular, by (7.17), we obtain that

$$\{ U\mu - U\nu \mid \mu, \nu \in M_{\mathcal{E}}^{+} \} \quad \text{and} \quad \{ U\mu \mid \mu \in M_E \}$$

are dense in $\mathcal{F}_e$.

Proof. We shall show that for any $\mu \in M_{\mathcal{E}}^{+}(A)$, we can find $\mu_n \in M_E^{+}(A)$, such that $U\mu_n \to U\mu$ in $(\mathcal{F}_e, \mathcal{E})$. Let K_n be a sequence of compact sets, $K_n \uparrow E$. Consider $\mu_n = I_{K_n}\mu$. Then $\mu_n \in M_E^{+}(A)$ and

$$\| U\mu - U\mu_n \|_{\mathcal{E}}^{2} = \mathcal{E}(U\mu, U\mu) - 2\mathcal{E}(U\mu, U\mu_n) + \mathcal{E}(U\mu_n, U\mu_n).$$

and $\mathcal{E}(U\mu_n, U\mu_n) = \int_{K_n} \widetilde{U\mu}_n d\mu$, where $\widetilde{U\mu}_n$ is a quasi-continuous modification (in the restricted sense) of $U\mu_n$, $\widetilde{U\mu}_n \geq 0$ and is quasi-continuous, so that $\widetilde{U\mu}_n \geq 0$ μ–a.e. (by [30], pg. 71). Hence,

$$
\begin{aligned}
\mathcal{E}(U\mu_n, U\mu_n) &= \int_{K_n} \widetilde{U\mu}_n d\mu \leq \int_E \widetilde{U\mu}_n d\mu \\
&= \mathcal{E}(U\mu, U\mu_n) = \int_E \widetilde{U\mu} \, d\mu_n \\
&\leq \int_E \widetilde{U\mu} \, d\mu = \mathcal{E}(U\mu, U\mu).
\end{aligned}
$$

Thus,

$$\| U\mu - U\mu_n \|_{\mathcal{E}}^{2} \leq 2 \left(\mathcal{E}(U\mu, U\mu) - \mathcal{E}(U\mu, U\mu_n) \right).$$

By the Monotone Convergence Theorem

$$\mathcal{E}(U\mu, U\mu_n) = \int_E \widetilde{U\mu} \, d\mu_n = \int_{K_n} \widetilde{U\mu} \, d\mu \to \int_E \widetilde{U\mu} \, d\mu = \mathcal{E}(U\mu, U\mu),$$

as $n \to \infty$, completing the proof. $\qquad\square$

We now consider the symmetric Markov (transient) process generated by a Dirichlet form [108]. Let $(\mathcal{F}_e, \mathcal{E})$ be a regular extended transient Dirichlet space and $M(E)$ be as defined earlier. Consider a zero mean Gaussian random field $\{ X_\mu, \ \mu \in M(E) \}$ with covariance $E(X_\mu X_\nu) = C(\mu, \nu)$ (associated with the Green's function) of the symmetric Markov process. We call it the $(\mathcal{F}_e, \mathcal{E})$–Gaussian random field defined on $(\Omega, \mathcal{F}, P)$.

Remark 7.3.2. *Röckner [108] considers the Gaussian random field $M_{\mathcal{E}}$ with covariance $\mathcal{E}(\mu, \nu)$, $\mu, \nu \in M_{\mathcal{E}}$. But Markov properties of the random field considered by Röckner are the same as of the field defined above in view of Lemma 7.3.1. We choose our approach because $M_{\mathcal{E}}$ may not be a vector space and our general theory with measure-indexed random fields involves the index set, which is a vector space satisfying conditions (A1), (A2), and (A3).*

Our first purpose is to produce a dual field for X_μ. For this, we observe that if $g \in \mathcal{F}_e \cap C_0(E)$, then there exists a sequence $\{\mu_n\}_{n=1}^\infty \subseteq M(E)$, such that $\|U\mu_n - g\|_\mathcal{E} \to 0$. Hence, X_{μ_n} is a Cauchy sequence in $L^2(\Omega, \mathcal{F}, P)$. Let us define

$$\hat{X}_g = \lim_{n\to\infty} X_{\mu_n} \quad g \in \mathcal{F}_e \cap C_0(E). \tag{7.18}$$

Then

$$E\left(\hat{X}_g X_\mu\right) = \mathcal{E}(g, U\mu) = \int_E \tilde{g}\, d\mu = \int_E g\, d\mu$$

for $\mu \in M(E)$. Let $G(E) = \mathcal{F}_e \cap C_0(E)$. Since $\mathcal{F}_e \cap C_0(E)$ is dense in $(\mathcal{F}_e, \mathcal{E})$, the Gaussian random field $\{\hat{X}_g, g \in G(E)\}$ is biorthogonal of the $(\mathcal{F}_e, \mathcal{E})$-Gaussian random field $\{X_\mu, \mu \in M(E)\}$, by Definition 7.2.1.

Exercise 7.3.1. *Prove that the Gaussian random fields $\{\hat{X}_g, g \in G(E)\}$ and $\{X_\mu, \mu \in M(E)\}$ are biorthogonal.*

We need the following lemma.

Lemma 7.3.2. *$G(E)$ has partition of unity property.*

Proof. Suppose $v \in \mathcal{F}_e \cap C_0(E)$ with $\mathrm{supp}(v) \subseteq G_1 \cup G_2$, where G_1, G_2 are open subsets of E. Then one can choose an open pre-compact set $\mathcal{O}$, such that

$$\mathrm{supp}(v) \setminus G_2 \subseteq \mathcal{O} \subseteq \overline{\mathcal{O}} \subseteq G_1.$$

By part (iii) of Lemma 7.4.1 (Appendix 7.4), there exists a function $\omega \in \mathcal{F}_e \cap C_0(E)$, such that $\omega = 1$ on $\mathcal{O}$ and $\mathrm{supp}(\omega) \subseteq G_1$. We can represent v as follows:

$$v = v\omega + (v - v\omega) = v_1 + v_2.$$

Then by part (i) of Lemma 7.4.1, $v_i \in \mathcal{F}_e \cap C_0(E)$ and $\mathrm{supp}(v_i) \subseteq G_i$, $i = 1, 2$. $\quad\square$

Definition 7.3.6. *A Dirichlet space $(\mathcal{F}_e, \mathcal{E}\}$ is called local if for every $u, v \in (\mathcal{F}_e, \mathcal{E}\}$, with $\mathrm{supp}(u\,dm)$, $\mathrm{supp}(v\,dm)$ being compact and disjoint, $\mathcal{E}(u, v) = 0$.*

As usual, denote by $K(C)$ the RKHS of $\{X_\mu\}_{\mu \in M(E)}$ and for $g \in G(E)$

$$f_g(\mu) = E\left(\hat{X}_g X_\mu\right).$$

We need the following lemma to show that the Markov property of the Gaussian random field $\{X_\mu, \mu \in M(E)\}$, whose covariance is $C(\mu, v) = \mathcal{E}(U\mu, Uv)$ is equivalent to the Dirichlet space $(\mathcal{F}_e, \mathcal{E}\}$ being *local*.

Lemma 7.3.3. *Let $K(C)$ be the RKHS of the covariance $C(\mu, v) = \mathcal{E}(U\mu, Uv)$, $\mu, v \in M(E)$. For the biorthogonal process $\hat{X}$ defined in (7.18) we define for any open subset $D \subseteq E$,*

$$\begin{aligned} M(D) &= \overline{\mathrm{span}}^{K(C)}\{f \mid f \in K(C), \mathrm{supp}(f) \subseteq D\} \\ \hat{M}(D) &= \overline{\mathrm{span}}^{K(C)}\{f_g \mid g \in K(C), \mathrm{supp}(g) \subseteq D\}. \end{aligned}$$

Then $M(D) = \hat{M}(D)$.

Proof. Since $f_g \in K(C)$ and $\operatorname{supp}(g) \subseteq D$ implies that $f_g \in M(D)$, it remains to prove $M(D) \subseteq \hat{M}(D)$. Let $f \in M(D)$ and denote $A = \operatorname{supp}(f)$. There exists an element $u \in \mathcal{F}_e$, such that

$$f(\mu) = E\left(\hat{X}_u X_\mu\right) = \mathcal{E}(U\mu, u) = \int_E u \, d\mu = 0$$

for any $\mu \in M(E)$ with $\operatorname{supp}(\mu) \subseteq A^c$. By Lemma 7.5.1,

$$u \in \left[\overline{\operatorname{span}}^{\mathcal{E}} \{U\mu \mid \mu \in M_E(A^c)\}\right]^\perp = \left(W_0^{A^c}\right)^\perp = \left(\mathcal{H}_0^{A^c}\right)^\perp = \mathcal{F}_A.$$

Using Lemma 7.5.2, there exists a sequence $\{g_n\}_{n=1}^\infty \subseteq \mathcal{F}_e \cap C_0(E)$ with $\operatorname{supp}(g_n) \subseteq D$, $n = 1, 2...$, and $g_n \to u$ in $(\mathcal{F}_e, \mathcal{E})$ as $n \to \infty$. Hence,

$$\hat{X}_{g_n} \to \pi(f) \text{ in } L^2(\Omega, \mathcal{F}, P),$$

where π is the isometry between $K(C)$ and $H(X)$. We conclude that

$$f_{g_n} = \int_E g_n \, d\mu \to f \text{ in } K(C).$$

Because $f_{g_n} \in \hat{M}(D)$, also the limit $f \in \hat{M}(D)$, giving the desired conclusion.
$\square$

Theorem 7.3.2. *Let $\left\{X_\mu, \mu \in M(E)\right\}$ be an $(\mathcal{F}_e, \mathcal{E})$-Gaussian random field with covariance $E\left(X_\mu X_\nu\right) = \mathcal{E}(\mu, \nu)$. Then X_μ has Germ Field Markov Property (GFMP) if and only if one of the following conditions holds true:*

(a) *$(\mathcal{F}_e, \mathcal{E})$ has local property,*

(b) *$\langle f_1, f_2 \rangle_{K(C)} = 0$ for any $f_1, f_2 \in K(C)$ with disjoint supports.*

Proof. The equivalence of GFMP of $\left\{X_\mu, \mu \in M(E)\right\}$ and condition (b) is a consequence of Lemmas 7.3.2 and 7.3.3 and Corollary 7.2.1. It suffices to prove the equivalence of (a) and (b). To show that (a) implies (b) we take $f_1, f_2 \in K(C)$ with $\operatorname{supp}(f_1) \cap \operatorname{supp}(f_2) = \varnothing$. Since $\overline{\operatorname{span}}^{\mathcal{E}} \{U\mu, \mu \in M(E)\} = \mathcal{F}_e$, there exist $u_1, u_2 \in \mathcal{F}_e$, such that $\langle f_1, f_2 \rangle_{K(C)} = \mathcal{E}(u_1, u_2)$ and

$$f_i(\mu) = \mathcal{E}(u_i, U\mu) = \int_E \tilde{u}_i \, d\mu, \; i = 1, 2, \; \mu \in M(E).$$

Let $A_i = \operatorname{supp}(f_i)$, $i = 1, 2$, then $\mathcal{E}(u_i, U\mu) = 0$ for $\mu \in M(E)$ with $\operatorname{supp}(\mu) \subseteq A^c$. By Lemma 7.5.1,

$$u_i \in \left[\overline{\operatorname{span}}^{\mathcal{E}} \{U\mu \mid \mu \in M_E(A_i^c)\}\right]^\perp = \left(W_0^{A_i^c}\right)^\perp = \left(\mathcal{H}_0^{A_i^c}\right)^\perp = \mathcal{F}_{A_i},$$

where $\mathcal{F}_{A_i} = \{v \in \mathcal{F}_e \mid \tilde{v} = 0 \text{ q.e. on } A_i^c\}$. Let D_i be a neighborhood of A_i, $i = 1, 2$, such that $D_1 \cap D_2 = \varnothing$. By Lemma 7.5.2 there exists sequences $\left\{g_n^i\right\}_{n=1}^\infty \subseteq \mathcal{F}_e \cap$

$C_0(E)$, with $\operatorname{supp}(g_n^i) \subseteq D_i$, $n = 1, 2\ldots$, and $g_n^i \to u$ in $(\mathcal{F}_e, \mathcal{E})$ as $n \to \infty$, $i = 1, 2$. Hence, because $\mathcal{E}$ is closed, and is local, we obtain

$$\langle f_1, f_2 \rangle_{K(C)} = \mathcal{E}(u_1, u_2) = \lim_{n \to \infty} \mathcal{E}(g_n^1, g_n^2) = 0.$$

To see that (b) implies (a) we let $v_1, v_2 \in \mathcal{F}_e \cap L^2(E, m)$ be such that $\operatorname{supp}(v_1 \, dm)$ and $\operatorname{supp}(v_2 \, dm)$ are disjoint. Since $K(C)$ and $(\mathcal{F}_e, \mathcal{E})$ are isometric, there exist $f_1, f_2 \in K(C)$, such that

$$\mathcal{E}(v_1, v_2) = \langle f_1, f_2 \rangle_{K(C)} \quad \text{and} \quad \mathcal{E}(v_i, U\mu) = f_i(\mu), \, i = 1, 2.$$

Let $A_i = \operatorname{supp}(v_i \, dm)$, $v_i = 0$ m–a.e. on A_i^c, giving $v_i = 0$ m–q.e. on A_i^c, $i = 1, 2$. This implies that $f_i(\mu) = 0$ for every $\mu \in M(E)$, with $\operatorname{supp}(\mu) \subseteq A_i^c$, and further, that $\operatorname{supp}(f_i) \subseteq A_i$, $i = 1, 2$, and we conclude that $\mathcal{E}(v_1, v_2) = 0$. $\qquad\square$

Exercise 7.3.2. *Show that conditions (a) and (b) of Theorem 7.3.2 are equivalent to the following condition:*

(b') $\langle f_1, f_2 \rangle_{K(C)} = 0$ *for* $f_1, f_2 \in K(C)$ *with compact and disjoint supports.*

As stated earlier in Lemma 7.3.1 and Remark 7.3.2, we obtain that for an open set $D \subseteq E$,

$$\Sigma(\overline{D}) \vee \Sigma(D^c) = \sigma\{X_\mu, \, \mu \in M(E)\} = F(E).$$

Then condition

$$F(D) \perp\!\!\!\perp F\left((\overline{D})^c\right) \big| F(\mathcal{O})$$

for all open sets $\mathcal{O} \supseteq \partial D$ is equivalent to GFMP on D.

Lemma 7.3.4. *For an open subset $D \subseteq E$,*

$$\sigma\{X_\mu, \operatorname{supp}(\mu) \subseteq D\} \vee \sigma\{X_\mu, \operatorname{supp}(\mu) \subseteq D^c\} = F(E),$$

equivalently, $F(D) \vee F(D^c) = F(E)$.

Proof. Let $\mu \in M(E)$, then $\mu = 1_D \mu + 1_{D^c} \mu$, giving

$$X_m u = X_{1_D \mu} + X_{1_{D^c} \mu},$$

so that $X_{1_{D^c} \mu}$ is $F(D)$-measurable. We take a sequence $(K_n)_{n=1}^\infty$ of compact subsets of E, such that $K_n \uparrow D$, then $|\mu|(D \setminus K_n) \to 0$. Similar as in the proof of Lemma 7.3.1 we can show that $\mathcal{E}(U\mu_n, U\mu_n) \to 0$, where $\mu_n = 1_{D \setminus K_n} \mu$. Hence,

$$U 1_{K_n} \mu \to U 1_D \mu \text{ as } n \to \infty \text{ in } (\mathcal{F}_e, \mathcal{E}),$$

giving

$$X_{1_{K_n} \mu} \to X_{1_D \mu} \text{ in } L^2(\Omega, \mathcal{F}, P).$$

This implies $X_{1_D \mu} \in F(D)$, completing the proof. $\qquad\square$

Corollary 7.3.1. *An $(\mathcal{F}_e, \mathcal{E})$-Gaussian random field $\{X_\mu, \mu \in M(E)\}$ has GFMP for open sets if and only if $(\mathcal{F}_e, \mathcal{E})$ has local property.*

Lemma 7.3.5. *For an $(\mathcal{F}_e, \mathcal{E})$-Gaussian random field $\{X_\mu, \mu \in M(E)\}$, we have*

(i) *if $A, B \subseteq E$ are closed subsets and $\mathcal{O} \subseteq E$ is open, then $B \subseteq A \cup \mathcal{O}$ implies*
$$F(A) \vee F(\mathcal{O}) = F(B),$$

(ii) *for any closed set A, $F(A) = \Sigma(A) = \bigcap_{\mathcal{O} \supseteq A} F(\mathcal{O})$, where the intersection is over open sets $\mathcal{O} \supseteq A$.*

Proof. We prove part (ii) only leaving the proof of (i) as an exercise (Exercise 7.3.3).

We can chose open sets U_n, $n = 1, 2, \ldots$, such that $U_n \supseteq \overline{U_{n+1}}$, and $\bigcap_{n=1}^{\infty} U_n = A$. We shall show that

$$\bigcap_{n=1}^{\infty} F\left(\overline{U_n}\right) = F(A),$$

this will imply $F(A) = \Sigma(A)$.

Denote $\mathcal{G} = \bigcap_{n=1}^{\infty} F\left(\overline{U_n}\right)$ and $\mathcal{F}_n = F\left(\overline{U_n}\right)$. For an integer k, let $\mu_1, \ldots, \mu_k \in M^+(E)$ and $Z = \prod_{i=1}^{k} X_{\mu_i}$. We shall show that $E(Z|, \mathcal{G})$ is $F(A)$ measurable, implying $\mathcal{G} \subseteq F(A)$, while the opposite inclusion is obvious. By Martingale Convergence Theorem, P-a.e.

$$E(Z|, \mathcal{G}) = \lim_{n \to \infty} E(Z|\mathcal{F}_n).$$

For $\mu \in M(E)$, let μ^n be Balayaged measure of $\overline{U_n}$ (see Appendix 7.5). Denote by X_{μ_n} the projection of X_μ on $\overline{\mathrm{span}}\{X_\nu, \nu \in M(E), \mathrm{supp}(\nu) \subseteq \overline{U_n}\}$ and write

$$X_{\mu - \mu^n} = X_\mu - X_{\mu^n}.$$

Now Z can be written as sums of the form

$$\prod_{i=1}^{k} \left(X_{\mu_i^n}\right)^{\alpha_i} \prod_{i=1}^{k} \left(X_{\mu_i - \mu_i^n}\right)^{\beta_i}, \quad \alpha_i, \beta_i \in \{0, 1\}, ; \sum_i (\alpha_i + \beta_i) = k.$$

Since $U\mu_i^n \in \mathcal{H}_0^{\overline{U_n}} = \mathcal{W}_0^{\overline{U_n}}$ and $U(\mu_i - \mu_i^n) \perp \mathcal{W}_0^{\overline{U_n}}$, $\prod_{i=1}^{k}\left(X_{\mu_i - \mu_i^n}\right)^{\beta_i}$ is independent of $\mathcal{F}_n$. Then

$$E(Z|\mathcal{F}_n) = \prod_{i=1}^{k} \left(X_{\mu_i^n}\right)^{\alpha_i} E\left(\prod_{i=1}^{k}\left(X_{\mu_i - \mu_i^n}\right)^{\beta_i}\right).$$

Using (3.3.18) of [30],

$$\mathcal{W}_0^A = \bigcap_{n=0}^{\infty} \mathcal{W}_0^{U_n} = \bigcap_{n=0}^{\infty} \mathcal{W}_0^{\overline{U_n}}.$$

Therefore, $P_{\mathcal{W}_0}^{\overline{U_n}}(U\mu_i) \to P_{\mathcal{W}_0}^A(U\mu_i)$, where $P_{\mathcal{W}_0}^{\overline{U_n}}$, $P_{\mathcal{W}_0}^A(U\mu_i)$ are projections, respectively, on $\overline{U_n}$, and A. This means that $X_{\mu_i^n} \to X_{\mu_i^A}$ in $L^2(\Omega, \mathcal{F}, P)$, where μ_i^A is the Balayage measure of μ on A. Thus, $\prod_{i=1}^k \left(X_{\mu_i^n}\right)^{\alpha_i} \to \prod_{i=1}^k \left(X_{\mu_i^A}\right)^{\alpha_i}$ in probability, and passing to a subsequence, if necessary, the convergence can be assumed to be P–a.s. Since $X_{\mu_i - \mu_i^n} \to X_{\mu_i^n - \mu_i^A}$ in $L^2(\Omega, \mathcal{F}, P)$ for $i = 1, 2, ..., k$, and $\left\{X_{\mu_i - \mu_i^n}, i = 1, 2, ..., k\right\}$ are joint Gaussian, then using the Fourier transform of the joint distribution, we obtain $\lim_{n \to \infty} E\left(\prod_{i=1}^k \left(X_{\mu_i - \mu_i^n}\right)^{\beta_i}\right)$ exists in $\mathbb{R}$; thus, $\lim_{n \to \infty} E(Z|\mathcal{F}_n)$ is $F(A)$-measurable because $\prod_{i=1}^k \left(X_{\mu_i^A}\right)^{\alpha_i}$ is $F(A)$ measurable, so $E(Z|\mathcal{F}_n)$ is $F(A)$ measurable. The polynomials in X_μ, $\mu \in M(E)$ are dense in $L^2(\Omega, \mathcal{F}, P)$, giving that $\mathcal{G} = F(A)$. $\square$

Exercise 7.3.3. *Prove part (i) of Lemma 7.3.5 using arguments similar to those in the proof of Lemma 7.3.4.*

Lemma 7.3.6. *Let $A \subseteq E$. The following are equivalent for an $(\mathcal{F}_e, \mathcal{E})$-Gaussian field $\left\{X_\mu, \mu \in M(E)\right\}$,*

(i) $F\left(\overline{A}\right) \perp\!\!\!\perp F\left(\overline{A^c}\right) | F(\partial A)$.

(ii) $F(A) \perp\!\!\!\perp F(A^c) | F(\partial A)$.

(iii) $F\left(\mathring{A}\right) \perp\!\!\!\perp F\left(\mathring{A^c}\right) | F(\partial A)$.

Here $\mathring{A}$ and $\mathring{A^c}$ denote the interiors of A and A^c, respectively.

Proof. The fact that (i) implies (iii) follows from $F\left(\mathring{A}\right) \subseteq F\left(\overline{A}\right)$ and $F\left(\mathring{A^c}\right) \subseteq F\left(\overline{A^c}\right)$.

To show that (iii) implies (i) we note that $\mathring{A} \cup \partial A = \overline{A}$ and $\mathring{A^c} \cup \partial A = \overline{A^c}$. By part (i) of Lemma 7.3.5 $F\left(\mathring{A}\right) \vee F(\partial A) = F\left(\overline{A}\right)$, and $F\left(\mathring{A^c}\right) \vee F(\partial A)$. The result is now a direct consequence of part (i) of Exercise 7.1.4.

Since $F\left(\mathring{A}\right) \subseteq F(A) \subseteq \left(\overline{A}\right)$ and $F\left(\mathring{A^c}\right) \subseteq F(A^c) \subseteq \left(\overline{A^c}\right)$, again by part (i) of Exercise 7.1.4, (i) $\Leftrightarrow$ (ii) $\Leftrightarrow$ (iii). $\square$

Lemma 7.3.7. *For an $(\mathcal{F}_e, \mathcal{E})$-Gaussian field $\left\{X_\mu, \mu \in M(E)\right\}$, the following are equivalent:*

(i) $F\left(\overline{\mathcal{O}}\right) \perp\!\!\!\perp F\left(\overline{\mathcal{O}^c}\right) | F(\partial\mathcal{O})$ *for all open subsets $\mathcal{O} \subseteq E$.*

(ii) $F\left(\overline{A}\right) \perp\!\!\!\perp F\left(\overline{A^c}\right) | F(\partial A)$ *for all subsets $A \subseteq E$.*

Proof. We only need to show that (i) implies (ii). By Lemma 7.3.6, it suffices to prove that for an arbitrary subset $A \subseteq E$,

$$\left(\mathring{A}\right) \perp\!\!\!\perp F\left(\mathring{A^c}\right) | F(\partial A). \tag{7.19}$$

Since the set $\mathring{A}$ is open, we have $F\left(\overline{\mathring{A}}\right) \perp\!\!\!\perp F\left(\overline{\mathring{A}^c}\right) | F(\partial\mathring{A})$. Now, to use part

(ii) of Exercise 7.1.4, we note that $\partial \mathring{A} \subseteq \partial A$, and since $\overline{\mathring{A}} \cup \overline{\mathring{A^c}} = E$, implying $F\left(\overline{\mathring{A}}\right) \vee F\left(\overline{\mathring{A^c}}\right) = F(E) \supseteq F(\partial A)$, we obtain

$$F\left(\overline{\mathring{A}}\right) \perp\!\!\!\perp F\left(\overline{\mathring{A^c}}\right) | F(\partial A).$$

An application of Exercise 7.1.4 yields (7.19).   □

We note that in Lemma 7.3.7, condition (i) is GFMP for all open sets, and condition (ii) is what we refer to as Markov property III for all sets. Combining Theorem 7.3.2, Corollary 7.3.1, and Lemma 7.3.7 we have the following result of Röckner ([108], Theorem 7.4).

Theorem 7.3.3. *For an $(\mathcal{F}_e, \mathcal{E})$-Gaussian field $\{X_\mu, \mu \in M(E)\}$, the following are equivalent:*

(i) X_μ *has the GFMP for all open sets.*

(ii) X_μ *has the Markov property III for all sets.*

(iii) $(\mathcal{F}_e, \mathcal{E})$ *has the local property.*

7.4   Appendix A: Dirichlet Forms, Capacity, and Quasi-Continuity

We recall some definitions and results from [30] needed in studying Gaussian random fields related to Dirichlet forms and the work of Dynkin [21] and Röckner [108].

Lemma 7.4.1 ([30] Theorem 1.4.2 and Lemma 1.4.2). *A regular extended (transient) Dirichlet space $(\mathcal{F}_e, \mathcal{E})$ has the following properties:*

(i) *If $u, v \in \mathcal{F}_e$, hen $u \vee v$, $u \wedge v$, $u \wedge 1$, u^+, u^- $\in \mathcal{F}_e$, and $uv \in \mathcal{F}_e \cap L^\infty(E, m)$ (the space of m–a.e. bounded real-valued functions on E) implies $uv \in (\mathcal{F}_e, \mathcal{E})$.*

(ii) *Let $\{u_n\}_{n=1}^\infty, u \in (\mathcal{F}_e, \mathcal{E})$ be such that $u_n \to u$ in $(\mathcal{F}_e, \mathcal{E})$ as $n \to \infty$. Let $\varphi(t)$ be a real-valued function with $\varphi(0) = 0$, $|\varphi(t) - \varphi(t')| \le |t - t'|$ for $t, t' \in \mathbb{R}$. Then $\varphi(u_n), u \in (\mathcal{F}_e, \mathcal{E})$, and $\varphi(u_n) \to \varphi(u)$, as $n \to \infty$, weakly with respect to $\mathcal{E}$. In addition, if $\varphi(u) = u$, then the convergence is strong with respect to the norm given by $\mathcal{E}$.*

(iii) *For $u \in C_0(E)$, there exists a sequence $\{u_n\}_{n=1}^\infty \subseteq \mathcal{F}_e \cap C_0(E)$, such that $u_n \to u$ uniformly.*

We need to define 0-order capacity to introduce the concept of *quasi-everywhere* and *quasi-continuity*. For $A \subseteq E$ denote

$$\mathcal{L}_A = \{u \,|\, u \in \mathcal{F}_e, u \ge 1 \, m-\text{a.e. on } A\}.$$

Definition 7.4.1. *The 0-order capacity, or simply capacity of an open subset $\mathcal{O} \subset E$ is defined as*

$$Cap_0(\mathcal{O}) = \begin{cases} \inf_{u \in \mathcal{L}_\mathcal{O}} \mathcal{E}(u, u) & \text{if } \mathcal{L}_\mathcal{O} \ne \varnothing \\ \infty & \text{otherwise} \end{cases}$$

For any subset $A \subseteq E$,

$$Cap_0(A) = \inf_{\mathcal{O} \supseteq A} Cap_0(\mathcal{A}),$$

where the infimum is taken over all open subsets $E \supseteq \mathcal{O} \supseteq A$.

We have the following properties of capacity.

Lemma 7.4.2 (Theorem 3.1.1 in [30]). *The capacity in Definition 7.4.1 has the following properties.*

(i) *If $A \subset B$, then $Cap_0(A) \leq Cap_0(B)$.*

(ii) *If $\{A_n\}_{n=1}^{\infty}$ is an increasing sequence of subsets of E, then $Cap\left(\bigcup_{n=1}^{\infty} A_n\right) = \sup_n Cap_0(A_n)$.*

(iii) *If $\{A_n\}_{n=1}^{\infty}$ is an decreasing sequence of compact subsets of E, then $Cap_0\left(\bigcap_{n=1}^{\infty} A_n\right) = \inf_n Cap_0(A_n)$.*

This is precisely the definition of Choquet's capacity.

Since for an open subset $\mathcal{O} \subseteq E$, $m(\mathcal{O}) \leq Cap_0(\mathcal{O})$, so that any set of capacity zero is m-negligible, the notion of capacity allows us to think of exceptional sets finer than m-negligible sets.

Definition 7.4.2. *A statement depending on $x \in A \subseteq E$ is said to hold quasi-everywhere, q.e., on A if there exists a subset $N \subseteq A$ of zero capacity, such that the statement is true for all $x \in A \setminus N$. A function u defined on E is called quasi-continuous if for any $\varepsilon > 0$, there exists an open set $\mathcal{O} \subseteq E$, such that $Cap_0(\mathcal{O}) < \varepsilon$ and the restriction of u to $E \setminus \mathcal{O}$, $u|_{E \setminus \mathcal{O}}$ is continuous.*

A function u on E is called quasi-continuous in a restricted sense if u is quasi-continuous on E_Δ, (i.e., with the above notation $u|_{E_\Delta \setminus \mathcal{O}}$ is continuous), where E_Δ is the one-point compactifcation of E and it is assumed that $u(\Delta) = 0$.

Given two functions u,v on E, the function v is said to be a quasi-continuous modification of u in the restricted sense, if v is quasi-continuous in the restricted sense and $v = u$, m-a.e.

Lemma 7.4.3 (Theorem 3.1.3 in [30]). *Every element $u \in \mathcal{F}_e$ admits a quasi-continuous modification in the restricted sense denoted by $\tilde{u}$.*

7.5 Appendix B: Balayage Measure

For any Borel set $A \subseteq E$, we define the following subspace of $(\mathcal{F}_e, \mathcal{E})$:

$$\mathcal{F}_{E \setminus A} = \{u \,|\, u \in \mathcal{F}_e, \tilde{u} = 0, \text{ q.e. on } A\},$$

and denote by $\mathcal{H}_0^A$ its orthogonal complement in $(\mathcal{F}_e, \mathcal{E})$, i.e. $\mathcal{H}_0^A = \mathcal{F}_{E \setminus A}^{\perp}$.

Definition 7.5.1. *For $v \in \mathcal{F}_e$ we define $S(v)$, the spectrum of v as the complement of the largest open set $\mathcal{O} \subseteq E$, such that $\mathcal{E}(v,u) = 0$ for any $u \in \mathcal{F}_e \cap C_0(E)$ with $\text{supp}(u) \subseteq \mathcal{O}$. In particular, for $\mu \in M_{\mathcal{E}}^+$, $\text{supp}(U\mu) = \text{supp}(\mu)$.*

We formulate the following lemma based on the statements on pg. 79 in [30].

Lemma 7.5.1. *Let A be an open or closed subset of E. Then*

$$\mathcal{H}_0^A = \overline{\operatorname{span}}^{\mathcal{E}} \left\{ U\mu \,|\, \mu \in M_E^+(A) \right\}-,$$

and (7.20)

$$\mathcal{H}_0^A = \overline{\operatorname{span}}^{\mathcal{E}} \left\{ v \in \mathcal{F}_e \,|\, S(V) \subseteq A \right\}.$$

Let for any Borel subset $A \subseteq E$, P_A denote the orthogonal projection on $\mathcal{H}_0^A$ in the Hilbert space $(\mathcal{F}_e, \mathcal{E})$. We know that for $\mu \in M_{\mathcal{E}}^+$ there corresponds a potential $u = U\mu \in \mathcal{F}_e$. Let $u_A = P_A u$, then $u_A \in U\left(M_{\mathcal{E}}^+\right)$, and hence $u_A = U\mu^A$, with $\mu^A \in M_{\mathcal{E}}^+$, $\operatorname{supp}\left(\mu^A\right) \subseteq \overline{A}$. Following [30] we call μ^A the Balayage measure (sweeping out) of μ on A. We need the following lemma.

Lemma 7.5.2. *Let A be a closed set and $A \subseteq D$, where $D \subseteq E$ is open. If*

$$u \in \mathcal{F}_A = \left\{ u \in \mathcal{F}_e \,|\, \tilde{u} = 0 \text{ q.e. on } E \setminus A \right\},$$

then there exists a sequence $\{g_n\}_{n=1}^{\infty} \subseteq \mathcal{F}_e \cup C_0(E)$ with $\operatorname{supp}(g_n) \subseteq D$ and

$$g_n \to u \text{ in } (\mathcal{F}_e, \mathcal{E}) \text{ as } n \to \infty. \tag{7.21}$$

Proof. If $u \in \mathcal{F}_A$, then by Lemma 7.4.1 $u^+, u^- \in \mathcal{F}_e$. Since $\widetilde{u^+} \leq |\tilde{u}| = |\tilde{u}|$ and $\widetilde{u^-} \leq |\tilde{u}| = |\tilde{u}|$, both u^+ and u^- are in $\mathcal{F}_e$. Without loss of generality, we can assume that u is non-negative and quasi-continuous.

Since $\mathcal{F}_e \cap C_0(E)$ is dense in $(\mathcal{F}_e, \mathcal{E})$, there exists a sequence $\{v_n\}_{n=1}^{\infty} \subseteq \mathcal{F}_e \cup C_0(E)$, such that $v_n \to u$ in $(\mathcal{F}_e, \mathcal{E})$. We can assume that $v_n \geq 0$ because we can always replace v_n by $v_n^+ = \frac{1}{2}v_n + \frac{1}{2}|v_n|$ and $v_n^+ \to u$ in $(\mathcal{F}_e, \mathcal{E})$ by part (ii) of Lemma 7.4.1. Let $h_n = v_n \wedge u = \frac{1}{2}(v_n + u) - \frac{1}{2}|v_n - u|$, again by Lemma 7.4.1 (ii), $h_n \to u$ in $(\mathcal{F}_e, \mathcal{E})$. Notice that h_n is bounded, $h_n \in \mathcal{F}_A$, and the closure of $\{x \,|\, x \in E, \, h_n(x) \neq 0\}$ is compact. We can choose a sequence $\{w_n'\}_{n=1}^{\infty} \subseteq C_0(E)$, $w_n' \geq 0$ and $w_n' \geq h_n$ q.e. with $\operatorname{supp}(w_n') \subseteq D$. Let us choose another sequence $\{w_n''\}_{n=1}^{\infty} \subseteq C_0(E)$, $w_n'' \geq 0$ with $\operatorname{supp}(w_n'') \subseteq D$ and $w_n'' \geq w_n' + 1$ for $x \in \operatorname{supp}(w_n')$. By Lemma 7.4.1 (iii), for each n we can find a subsequence $\{w_m^n\}_{m=1}^{\infty} \subseteq (\mathcal{F}_e, \mathcal{E}) \cap C_0(E)$, $w_m^n \geq 0$ with $\operatorname{supp}(w_m^n) \subseteq \{x \,|\, w_n''(x) \neq 0\}$, and such that $\|w_m^n - w_n''\|_{C_0(E)} \to 0$ as $m \to \infty$. Consequently, for each n we can find $w_n \in (\mathcal{F}_e, \mathcal{E}) \cap C_0(E)$, such that $w_n \geq 0$ and $w_n(x) \geq w_n''(x)$ for all x, then $w_n \geq h_n$ q.e. Now for any n, we select a sequence $u_m^n \subseteq (\mathcal{F}_e, \mathcal{E}) \cap C_0(E)$, such that $u_m^n \to v_n'$ as $m \to \infty$ in $(\mathcal{F}_e, \mathcal{E})$. Let $e_m^n = w_n \wedge u_m^n = \frac{1}{2}(w_n - u_m^n) - \frac{1}{2}|w_n - u_m^n|$ and using Lemma 7.4.1 (ii) again we can show that

$$e_m^n \to v_n' \text{ in } (\mathcal{F}_e, \mathcal{E}) \text{ as } m \to \infty.$$

Notice that $\operatorname{supp}(e_m^n) \subseteq D$ and $v_n' \to u$ in $(\mathcal{F}_e, \mathcal{E})$ as $n \to \infty$. Therefore we can find a sequence $\{g_n\}_{n=1}^{\infty} \subseteq (\mathcal{F}_e, \mathcal{E}) \cap C_0(E)$, such that $\operatorname{supp}(g_n) \subseteq D$ and

$$g_n \to u \text{ in } (\mathcal{F}_e, \mathcal{E}) \text{ as } n \to \infty.$$

$\square$

7.6 Appendix C: Example

An interesting example of a Markov field where Theorem 7.3.3 can be used is the de Wijs process occurring in geostatistical literature [85]. For the various applications we refer the reader to the work of Mondal [89]. The de Wijs process is a Gaussian field with generalized covariance

$$C(\mu,\nu) = - \iint_{\mathbb{R}^2} \log \|x-y\|_{\mathbb{R}^2} \, \mu(dx)\,\nu(dy),$$

where $\mu,\nu \in M\left(\mathbb{R}^2\right)$ are non-atomic measures with the total zero mass. Using a result of Port and Stone ([106], pg. 70) with the transition probability

$$p_t(x,y) = \frac{1}{2\pi t} exp\left\{-\frac{1}{2t}\|x-y\|_{\mathbb{R}^2}\right\},$$

we obtain

$$C(\mu,\nu) = \iint_{\mathbb{R}^2}\left(\int_0^{\infty} p_t(x,y)\,dt\right)\mu(dx)\,\nu(dy).$$

As $\|x-y\|_{\mathbb{R}^2}$ is negative definite, $p_t(x,y)$ is positive definite in x,y by Theorem 1.1.1 and hence, we conclude that $C(\mu,\nu)$ is a positive definite function of μ and ν. Let us consider the Gaussian random field $\left\{X_\mu,\ \mu \in M\left(\mathbb{R}^2\right)\right\}$. The above result of [106] shows that it is associated with Green's function of a two-dimensional Brownian motion, which is clearly Markov. Hence, from our results (Lemma 7.3.2, Lemma 7.3.3, and Theorem 7.2.1) the Wijs process has GFMP.

Chapter 8

Markov Property of Gaussian Fields and Dirichlet Forms

Using the results of Chapter 7, we now derive the Markov Property for Gaussian random field $\{\xi_t, t \in T\}$, where T is an open subset of $\mathbb{R}^n$. As a consequence, one can obtain the results of Künch [67] and Pitt [104]. In the second part of this chapter, we show the connection between Gaussian Markov Fields (GMF) and biorthogonal fields and Dirichlet forms. We assume as in Chapter 7 that the Gaussian fields are defined on a complete probability space $(\Omega, \mathcal{F}, P)$.

8.1 Markov Property for Ordinary Gaussian Random Fields

Let E be a separable locally compact Hausdorff space and $\{\xi_t, t \in E\}$ be a centered Gaussian random field. Then the Markov property of ξ_t can be derived from our general framework. In particular we derive the work of Künch [66] and Pitt [104]. We first consider the case when E is an open domain $T \subseteq \mathbb{R}^n$.

Let $A \subseteq T$ and, as in Chapter 7, we denote by $\overline{A}$ the closure of A in T and by ∂A the boundary of A in T. Let for any $B \subseteq T$, $\mathcal{F}(B) = \sigma\{\xi_t, t \in B\}$.

Definition 8.1.1. *A centered Gaussian random field $\{\xi_t, t \in T\}$ has the simple Markov Property on a subset $A \subseteq T$, where T is an open subset of $\mathbb{R}^n$, if*

$$\mathcal{F}(\overline{A}) \perp\!\!\!\perp \mathcal{F}(A^c) \,|\, \mathcal{F}(\partial A)$$

Lévy observed that his multiparameter Brownian motion does not have this Markov Property, so he proposed in [73] the following definition introduced as condition (8.1). We state it directly for a centered Gaussian random field $\{\xi_t, t \in T\}$.

Definition 8.1.2. *A centered Gaussian random field $\{\xi_t, t \in T\}$ has Markov Property on a subset $A \subseteq E$ if*

$$\mathcal{F}(\overline{A}) \perp\!\!\!\perp \mathcal{F}(\overline{A}^c) \,|\, \Sigma(\partial A), \tag{8.1}$$

where for any set $B \subseteq \mathbb{R}^n$, $\Sigma(B) = \cap_{\mathcal{O} \subseteq B} \mathcal{F}(\mathcal{O})$ and the intersection is over open sets $\mathcal{O} \subseteq \mathbb{R}^n$.

Using results on conditional independence from Section 7.1, one can obtain the following lemma:

Lemma 8.1.1. *The following are equivalent for a stochastic random field* $\{\xi_t, t \in E\}$ *and a subset* $A \subseteq E$

(i) $\quad \mathcal{F}(\overline{A}) \perp\!\!\!\perp \mathcal{F}\left(\overline{A}^c\right) | \Sigma(\partial A)$

(ii) *For every open set* $\mathcal{O} \supseteq \partial A$, $\mathcal{F}(A) \perp\!\!\!\perp \mathcal{F}\left(\overline{A}^c\right) | \mathcal{F}(\mathcal{O})$

(iii) $\Sigma(\overline{A}) \perp\!\!\!\perp \Sigma\left(\overline{A}^c\right) | \Sigma(\partial A)$ *(this condition is the GFMP)*

Exercise 8.1.1. *Prove Lemma 8.1.1.*

We assume that the covariance function of the Gaussian random field ξ_t is continuous, that is, $(t,s) \to E\left(\xi_t \xi_s\right) = C_\xi(t,s)$ is continuous in (t,s), or equivalently, $T \ni t \to \xi_t \in L^2(\Omega, \mathcal{F}, P)$ is continuous. Let $M(T) = \{\varphi\, dt, \; \varphi \in C_0^\infty(T)\}$. We know from Example 7.1.1 that $M(T)$ satisfies assumptions (A1)–(A3) of Section 7.1. We associate with $\{\xi_t, \; t \in T\}$ a generalized Gaussian random field $\{X_\varphi, \; \varphi \in C_0^\infty(T)\}$ defined by

$$X_\varphi = \int_T \xi_t \varphi(t)\, dt, \; \varphi \in C_0^\infty(T) \tag{8.2}$$

Lemma 8.1.2. *For any open set* $\mathcal{O} \subseteq T$,

$$H(X : \mathcal{O}) = H(\xi : \mathcal{O}),$$

where $H(X : \mathcal{O}) = \overline{\text{span}}\{X_\varphi, \; \varphi \in C_0^\infty(T), \; \text{supp}(\varphi) \subseteq \mathcal{O}\}$, *and* $H(\xi : \mathcal{O}) = \overline{\text{span}}\{\xi_t, \; t \in \mathcal{O}\}$.

Proof. Clearly $H(X : \mathcal{O}) \subseteq H(\xi : \mathcal{O})$. To show the converse, for any $t_0 \in O$ we choose a sufficiently small $\varepsilon_0 > 0$, such that $\{t, \; |t - t_0| < \varepsilon_0\} \subseteq \mathcal{O}$. For $\varepsilon < \varepsilon_0$ let

$$\varphi_\varepsilon(t) = \begin{cases} \varepsilon^{-n} c_n \exp\left\{-\dfrac{\varepsilon^2}{\varepsilon^2 - |t - t_0|^2}\right\}, & |t - t_0| < \varepsilon \\ 0 & \text{otherwise} \end{cases}$$

be the Friedrichs' mollifier. The constant c_n is chosen so that $\int_T \varphi_\varepsilon(t)\, dt = 1$. Clearly $\text{supp}(\varphi_\varepsilon) \subseteq \mathcal{O}$ for $\varepsilon < \varepsilon_0$. It is well known that $\varphi_\varepsilon \in C_0^\infty(T)$ and

$$\lim_{\varepsilon \to 0} \left| \int_{\mathbb{R}^n} \xi_t \varphi_\varepsilon(t)\, dt - \xi_{t_0} \right| \to 0,$$

giving $\xi_{t_0} \in H(X : \mathcal{O})$, which completes the proof. $\qquad\qquad \square$

Corollary 8.1.1. *Let* $\{X_\varphi, \; \varphi \in C_0^\infty(T)\}$ *be as in (8.2). Then for any open set* $\mathcal{O}$, X_φ *has Markov Property on* $\mathcal{O}$ *if and only if* X_φ *has GFMP on* $\mathcal{O}$.

Let $K(C_X)$ and $K(C_\xi)$ be the reproducing kernel Hilbert spaces of $\{X_\varphi, \; \varphi \in C_0^\infty(T)\}$ and $\{\xi_t, \; t \in T\}$, respectively, and we denote by $H(X)$ and $H(\xi)$ the linear subspaces of $L_2(\Omega, F, P)$ generated by $\{X_\varphi, \; \varphi \in C_0^\infty(T)\}$ and $\{\xi_t, \; t \in T\}$, respectively. Notice that since $C_\xi(s,t)$ is continuous, every element in $K(C_\xi)$ is also a continuous function on T. Recall the stochastic integral defined in Definition 2.2.3 and consider the inverse mappings π_X^{-1} and π_ξ^{-1}

$$\pi_X^{-1} : H(X) \to K(C_X), \quad \left(\pi_X^{-1}(Y)\right)(\varphi) = E\left(YX_\varphi\right), \quad \text{for } Y \in H(X),$$

$$\pi_\xi^{-1} : H(\xi) \to K(C_\xi), \quad \left(\pi_\xi^{-1}(Y)\right)(t) = E\left(Y\xi_t\right), \quad \text{for } Y \in H(\xi).$$

We know from Chapter 2 that both mappings π_X^{-1} and π_ξ^{-1} are isometries. Since by Lemma 8.1.2, $H(X) = H(\xi)$, the map $J = \pi_X^{-1}\pi_\xi$ is an isometry between RKHSs $K(C_\xi)$ and $K(C_X)$. We can explicitly express the map J. If $f \in K(C_\xi)$, then

$$
\begin{aligned}
(Jf)(\varphi) &= \left(\pi_X^{-1}\left(\pi_\xi(f)\right)\right)(\varphi) = E\left(\left(\pi_\xi(f)\right)X_\varphi\right) \\
&= E\left(\left(\pi_\xi(f)\right)\int_T \xi_t\varphi(t)\,dt\right) \\
&= \int_T E\left(\left(\pi_\xi(f)\right)\xi_t\right)\varphi(t)\,dt \\
&= \int_T f(t)\varphi(t)\,dt. \tag{8.3}
\end{aligned}
$$

Note that (8.3) defines a linear functional on $C_0^\infty(T)$. We need to define the support of a linear functional on $C_0^\infty(T)$ similar as in Definition 7.1.1 and Example 7.1.1.

Definition 8.1.3. *Let L be a linear functional on $C_0^\infty(T)$. We define* $\text{supp}(L)$ *as the complement of the largest open set $\mathcal{O} \subseteq T$, such that $L(\varphi) = 0$ for all $\varphi \in C_0^\infty(T)$ with $\text{supp}(\varphi) \subseteq \mathcal{O}$ (as usual $\text{supp}(f)$ is defined as the complement in T of the largest open set $\mathcal{O} \subseteq T$ such that $f(t) = 0$ on $\mathcal{O}$).*

Exercise 8.1.2. *Verify that for f_1, f_2 and $f \in K(C_\xi)$,*

$$\text{supp}(Jf) = \text{supp}(f) \tag{8.4}$$

and

$$\langle Jf_1, Jf_2 \rangle_{K(C_X)} = \langle f_1, f_2 \rangle_{K(C_\xi)}. \tag{8.5}$$

Now we can state and prove the following improvement of a result in Künsch [66] and Pitt [104].

Theorem 8.1.1. *Let T be an open set of $\mathbb{R}^n$ and $\{\xi_t, t \in T\}$ be a centered Gaussian process with continuous covariance. Then ξ_t has Markov Property for all open subsets of T if and only if the following hold.*

(a) *If $f_1, f_2 \in K(C_\xi)$ and $\text{supp}(f_1) \cap \text{supp}(f_2) = \varnothing$, then $\langle f_1, f_2 \rangle_{K(C_\xi)} = 0$.*

(b) *If $f \in K(C_\xi)$ and $f = f_1 + f_2$, where f_1 and f_2 are continuous functions on T having disjoint supports, then $f_1, f_2 \in K(C_\xi)$.*

Proof. We know that the condition that $\{\xi_t, t \in T\}$ has Markov Property for all open subsets of T is equivalent to the condition that $\{X_\varphi, \varphi \in C_0^\infty(T)\}$ has GFMP, where X_φ is defined in (8.2). Hence, we only need to show that conditions (a) and (b) of Theorem 7.2.1 are equivalent to conditions (a) and (b) of this theorem. It is easy to see that conditions (a) of both theorems are equivalent because of (8.4) and (8.5). To show that (b) of Theorem 7.2.1 implies (b) of this theorem, we let $f \in K(C_\xi)$ and assume that $f = f_1 + f_2$ with f_1, f_2 being continuous and having disjoint supports. Then for any $\varphi \in C_0^\infty(T)$

$$(Jf)(\varphi) = \int_T f_1 \varphi \, dt + \int_T f_2 \varphi \, dt = F_1(\varphi) + F_2(\varphi).$$

Since F_1 and F_2 are linear functionals on $C_0^\infty(T)$ and $\mathrm{supp}(F_i) = \mathrm{supp}(f_i)$, $i = 1, 2$, we conclude that $F_i \in K(C_X)$ by (b) of Theorem 7.2.1. Then $F_i(\varphi) = \int_T f_i' \varphi \, dt$ with $f_i' \in K(C_\xi)$, $i = 1, 2$. Now

$$\int_T f_i \varphi \, dt = \int_T f_i' \varphi \, dt \quad i = 1, 2,$$

for all $\varphi \in C_0^\infty(T)$, which implies that $f_i = f_i'$, $i = 1, 2$, because f_i and f_i' are continuous. Hence $f_i \in K(C_\xi)$.

Conversely, if $F \in K(C_X)$, $F = F_1 + F_2$, with F_1 and F_2 being linear functionals on $C_0^\infty(T)$ and having disjoint supports, then $F(\varphi) = \int_T f \varphi \, dt$ with $f \in K(C_\xi)$. Since $\mathrm{supp}(f) = \mathrm{supp}(F) \subseteq \mathrm{supp}(F_1) \cup \mathrm{supp}(F_2)$, we can define for $i = 1, 2$

$$f_i(t) = \begin{cases} f(t) & t \in \mathrm{supp}(F_i) \\ 0 & \text{otherwise} \end{cases}$$

Then, f_1 and f_2 are continuous and $\mathrm{supp}(f_i) \subseteq \mathrm{supp}(F_i)$, $i = 1, 2$. Furthermore, $f = f_1 + f_2$. By (b) of this theorem $f_i \in K(C_\xi)$. Then

$$F_1(\varphi) - \int_T f_1 \varphi \, dt = \int_T f_2 \varphi \, dt - F_2(\varphi), \quad \varphi \in C_0^\infty(T)$$

Both sides are linear functionals on $C_0^\infty(T)$ and they have disjoint supports, so they are zero functionals by Lemma 7.1.1 (b), hence $F_i(\varphi) = \int f_i \varphi \, dt$, and $F_i \in K(C_X)$, $i = 1, 2$. $\qquad\square$

We notice from the proof of Theorem 8.1.1 that the choice of the space of measures $M(T)$ is not unique. Any space of measures $M(T)$ which satisfies the following additional assumptions (a) and (b) besides (A1)–(A3) of Section 7.1 will lead to the conclusions of Theorem 8.1.1.

(a) If f is a continuous function on T and for every open subset $\mathcal{O} \subseteq T$, $\int_T f \, d\mu = 0$ for all $\mu \in M(T)$ with $\mathrm{supp}(\mu) \subseteq \mathcal{O}$, then $f = 0$ on $\mathcal{O}$.

(b) For every open subset $\mathcal{O} \subseteq T$, $H(\xi : \mathcal{O}) = H(X : \mathcal{O}, M(T))$, where $X_\mu = \int_T \xi_t \, d\mu$, $H(X : \mathcal{O}, M(T)) = \overline{\mathrm{span}}\{X_\mu, \mu \in M(T) \text{ with } \mathrm{supp}(\mu) \subseteq \mathcal{O}\}$.

This observation allows for a generalization of T from a subset of $\mathbb{R}^n$ to any separable locally compact Hausdorff space. Let E be a separable locally compact Hausdorff space and $\{\xi_t, t \in E\}$ be a centered Gaussian process with continuous covariance function. Let m be a positive Radon measure on E with $\text{supp}(m) = E$. Then we define a space of measures

$$M(E) = \left\{ \sum_{i=1}^{n} 1_{A_i} \varphi_i \, dm \, \middle| \, n \geq 1, A_i \in \mathcal{B}(E), \varphi_i \in C_0^\infty(E) \right\},$$

where $C_0^\infty(E)$ denotes as usual the space of all continuous functions on E with compact support. By Example 7.1.1, $M(E)$ satisfies assumptions (A1)–(A3) of Section 7.1. Also, if f is continuous on E and $\int_E f \, d\mu = 0$ for all $\mu \in M(E)$ with $\text{supp}(\mu) \subseteq \mathcal{O}$, then $\int_E f\varphi \, dm = 0$ for all $\varphi \in C_0^\infty(E)$ with $\text{supp}(\varphi) \subseteq \mathcal{O}$. This implies that $f = 0$ on $\mathcal{O}$ because f is continuous and $\text{supp}(m) = E$.

If we define a measure indexed centered Gaussian field

$$X_\mu = \int_E \xi_t \, \mu(dt), \quad \mu \in M(E),$$

then we can show the following lemma.

Lemma 8.1.3. *For every open subset $\mathcal{O} \subseteq E$, $H(\xi : \mathcal{O}) = H(X : \mathcal{O}, M(E))$.*

Proof. It is enough to show the lemma for a precompact open subset $\mathcal{O}$. Let $\mu \in M(E)$ with $\text{supp}(\mu) \subseteq \mathcal{O}$. Any continuous function f on E can be approximated in the topology of pointwise convergence on the subset $\mathcal{O}$ by functions f_n of the form

$$f_n(t) = \sum_{i=1}^{m_n} f(t_i) 1_{A_i}(t)$$

with $t_i \in \mathcal{O}$ and $A_i \in \mathcal{B}(E)$, that is, $f_n \to f$ pointwise on $\mathcal{O}$. Let

$$\xi_n(t) = \sum_{i=1}^{m_n} \xi_{t_i} 1_{A_i}(t), \quad \text{with } t_i \in \mathcal{O}.$$

Then $\xi_n(t) \to \xi_t$ for every $t \in \mathcal{O}$ in $L_2(\Omega, F, P)$. Then $\int_E \xi_n(t) d\mu \to \int_E \xi_t d\mu = X_\mu \in H(\xi : \mathcal{O})$.

On the other hand, let $t_0 \in \mathcal{O}$. Then we can choose a precompact open subset $\mathcal{O}_n$, with $t_0 \in \mathcal{O}_n$, such that $\overline{\mathcal{O}_n} \subseteq \mathcal{O}$ and $\overline{\mathcal{O}_n} \downarrow \{t_0\}$. Let $d\mu_n = \alpha_n 1_{\mathcal{O}_n} dm$ with $\alpha_n = 1/m(\mathcal{O}_n)$. Then we can see that $\mu_n \in M(E)$ with $\text{supp}(\mu_n) \subseteq \mathcal{O}$ and

$$E \left| \int_E \xi_t d\mu_n - \xi_{t_0} \right| \leq E \int_{\mathcal{O}_n} \alpha_n |\xi_t - \xi_{t_0}| \, dm$$

$$- \alpha_n \int_E 1_{\mathcal{O}_n} E |\xi_t - \xi_{t_0}| \, dm$$

$$\leq \sup_{t \in \mathcal{O}_n} E |\xi_t - \xi_{t_0}| \alpha_n \int_{\mathcal{O}_n} dm$$

$$= \sup_{t \in \mathcal{O}_n} E \left| \xi_t - \xi_{t_0} \right|.$$

Since all subsets $\mathcal{O}_n$ are precompact and contained in a compact set $\overline{\mathcal{O}_1}$, and $\mathcal{O}_n \downarrow \{t_0\}$, by uniform continuity of $E|\xi_t - \xi_{t_0}|$ on O_1 we have $\sup_{t \in \mathcal{O}_n} E|\xi_t - \xi_{t_0}| \to 0$ as $n \to \infty$. Hence, $\xi_{t_0} \in H(X : \mathcal{O}, M(E))$. $\square$

We shall have the following isometry between Reproducing Kernel Hilbert Spaces $J : K\left(C_\xi\right) \to K\left(C_X\right)$,

$$(Jf)(\mu) = \int_E f \, d\mu, \quad f \in K\left(C_\xi\right).$$

Similar to (8.4) and (8.5), for $f \in K\left(C_\xi\right)$, we have

$$\mathrm{supp}(Jf) = \mathrm{supp}(f) \tag{8.6}$$

and

$$\langle Jf_1, Jf_2 \rangle_{K(C_X)} = \langle f_1, f_2 \rangle_{K\left(C_\xi\right)}. \tag{8.7}$$

The following theorem is an extension of Theorem 8.1.1 and the proof is almost the same and hence omitted.

Theorem 8.1.2. *Let E be a separable locally compact Hausdorff space, $\{\xi_t, t \in E\}$ be a centered Gaussian process with continuous covariance and $K\left(C_\xi\right)$ be the RHKS of its covariance. Then ξ_t has the Markov Property for all open sets if and only if*

(a) *If $f_1, f_2 \in K\left(C_\xi\right)$ have disjoint supports, then $\langle f_1, f_2 \rangle_{K\left(C_\xi\right)} = 0$.*

(b) *If $f \in K\left(C_\xi\right)$ and $f = f_1 + f_2$, where f_1 and f_2 are continuous and have disjoint supports, then $f_i \in K\left(C_\xi\right), i = 1, 2$.*

We also obtain the following theorem similar to Theorem 7.2.2.

Theorem 8.1.3. *Let E be a separable locally compact Hausdorff space, $\{\xi_t, t \in E\}$ be a centered Gaussian process with continuous covariance and $K\left(C_\xi\right)$ be the RHKS of its covariance. Then ξ_t has the Markov Property for all pre-compact open sets if and only if*

(a) *If $f_1, f_2 \in K\left(C_\xi\right)$ have disjoint supports and at least one of the supports is compact, then $\langle f_1, f_2 \rangle_{K\left(C_\xi\right)} = 0$.*

(b) *If $f \in K\left(C_\xi\right)$, $f = f_1 + f_2$ with f_1 and f_2 being continuous and having disjoint supports of which at least one is compact, then $f_i \in K\left(C_\xi\right), i = 1, 2$.*

Remark 8.1.1. *We do not want to introduce the concept of ultradistributions [60], [59], [61] in this book. If one studies Gaussian processes indexed by ultradistributions, one can derive the results for GFMP for stationary processes in terms of their spectral measure giving interesting results of*

Kotani [62], Kotani and Okabe [63], and Pitt [105]. It is also possible to obtain the result of Kusuoka [69] (in Gaussian case). For details, the reader is referred to [83].

8.2 Gaussian Markov Fields and Dirichlet Forms

We first consider measure indexed Gaussian random fields. Let E be a second-countable locally compact Hausdorff space and $X = \{X_\mu, \mu \in M(E)\}$ be a centered Gaussian random field with $M(E)$, the space of Radon signed measures on E satisfying conditions (A1)-(A2) of Section 7.1. Let $\hat{X} = \{\hat{X}_g, g \in G(E)\}$ be another Gaussian random field defined on the same probability space as X, where $G(E)$ is a dense subspace of $C_0(E)$, the space of continuous functions with compact support in E. Let us assume that $\hat{X}$ is a biorthogonal dual random field of X, that is $H(X) = H(\hat{X})$ and

$$E\hat{X}_g X_\mu = \int_E g\, d\mu, \quad g \in G(E) \text{ and } \mu \in M(E).$$

Hence, $f_g(\mu) = \int_E g\, d\mu$ is an element of $K(C_X)$, the RKHS of the covariance C_X of X. We introduced the following two spaces in Section 7.2, to study the GFMP of the Gaussian random field X,

$$
\begin{aligned}
M(D) &= \overline{\text{span}}\{f \mid f \in K(C_X), \text{supp}(f) \subseteq D\} \\
\hat{M}(D) &= \overline{\text{span}}\{f_g \mid \text{supp}(g) \subseteq D\}
\end{aligned}
$$

where D is an open subset of E.

We assume that for any open subset $D \subseteq E$, $H(X:D) \vee H(X:D^c) = H(X)$ and $M(D) = \hat{M}(D)$. By Theorem 7.2.3, a Gaussian random field X has GFMP if and only if

$$\hat{M}(D_1) \perp \hat{M}(D_2) \quad \text{if } D_1 \cap D_2 = 0 \text{ for all open subsets } D_1, D_2 \subseteq E. \quad (8.8)$$

Let us now consider the dual process $\hat{X}$ of the Gaussian Markov field X satisfying condition (8.8). Then its covariance

$$C_{\hat{X}}(g,g') = E(\hat{X}_g \hat{X}_{g'}), \quad g,g' \in G(E)$$

can be extended as a continuous positive bilinear form to $C_0(E) \otimes C_0(E)$. Hence by the Riesz representation theorem, there exists a positive measure v on $(E \times E, \mathcal{B}(E) \otimes \mathcal{B}(E))$ such that

$$C_{\hat{X}}(g,g') = \int_{E \times E} g(x)g'(y)\, v(dx, dy).$$

Since X has GFMP, then, as stated above, $\hat{M}(D_1) \perp \hat{M}(D_2)$ gives that

$$\mathcal{E}(g,g') = C_{\hat{X}}(g,g') = \int_{\Delta_E} g(x)g'(y)\, v_E(dx, dy)$$

where $v_E = v$ on the diagonal Δ_E of $E \times E$. Let $m(A) = v_E(\Delta_A)$ be the reference measure and let the reference function be equal to one on E; see Definition 7.3.4. We now extend $\mathcal{E}$ to a Dirichlet form on $\mathcal{F}_e = L^2(E,m)$ (see Appendix 7.4).

Exercise 8.2.1. *Show that conditions $(\mathcal{F}_e\text{-}1)$-$(\mathcal{F}_e\text{-}4)$ are satisfied for $(\mathcal{F}_e,\mathcal{E})$.*

Observe that since $\hat{X}$ is a dual process of X, we have for all $g \in C_0(E)$ and $\mu \in M(E)$,

$$
\begin{aligned}
\int_E |g|d|\mu| &= E\hat{X}_{|g|}X_{|\mu|} \\
&\leq \left(E\hat{X}_{|g|}^2\right)^{1/2}\left(EX_{|\mu|}^2\right)^{1/2} \\
&\leq \left(\int_{\Delta_E}|g(u)g(v)|dv_E(u,v)\right)^{1/2}C^{1/2}(|\mu|,|\mu|) \\
&\leq \mathcal{E}^{1/2}(g,g)C^{1/2}(|\mu|,|\mu|).
\end{aligned}
$$

Hence μ is a measure of finite energy (Example 7.1.2).

Note that for any $\mu \in M^+(E)$ which is of finite energy, and for a quasi-continuous modification (in restricted sense) $\tilde{g}$ of $g \in C_0(E)$, by Theorem 3.2.2 in [30], there exists a unique element $U_\mu \in C_0(E)$ (the potential of μ) such that

$$
E\hat{X}_{U_\mu}\hat{X}_g = \int_E \tilde{g}d\mu = EX_\mu\hat{X}_g.
$$

Since $H(X) = H(\hat{X})$, we can see that

$$
X_\mu = \hat{X}_{U_\mu} \quad \text{for all } \mu \in M^+(E).
$$

By linearity $\hat{X}_{U_\mu} = X_\mu$ for all $\mu \in M(E)$, that is, with $\mu \in M(E)$, we consider $\mu^\pm \in M^+(E)$, then $\mu = \mu^+ - \mu^-$, and $U_\mu = U_{\mu^+} - U_{\mu^-}$. Thus

$$
C_X(\mu,\mu') = \mathcal{E}(U_\mu,U_{\mu'}),
$$

that is, the covariance of $\{X_\mu, \ \mu \in M(E)\}$ is given by the potential associated with the Dirichlet form $\mathcal{E}$. Thus, we have the following theorem.

Theorem 8.2.1. *A Gaussian random field $\{X_\mu, \ \mu \in M(E)\}$ having a dual process $\{\hat{X}_g, \ g \in C_0(E)\}$ satisfying $\hat{M}(D) = M(D)$ for all open subsets $D \subseteq E$ has GFMP on all open subsets of E if and only if there exists a Dirichlet form $(\mathcal{E},\mathcal{D}(\mathcal{E}))$ on $C_0(E)$ such that $C_X(\mu,\mu') = \mathcal{E}(U_\mu,U_{\mu'})$ for all $\mu,\mu' \in M(E)$.*

Remark 8.2.1. *On a complete probability space $(\Omega,\mathcal{F},P)$ consider $\mathcal{N} = \{A \subseteq \mathcal{F}|P(A) = 0\}$. Let $\mathcal{A},\mathcal{B},\Sigma$ be sub σ-fields of $\mathcal{F}$ containing $\mathcal{N}$. Σ is called a splitting σ-field for $\mathcal{A}$ and $\mathcal{B}$ if $\mathcal{A} \perp\!\!\!\perp \mathcal{B}|\Sigma$.*

The spitting field $\Sigma_X(\partial D)$ is generated by the values of the Markov process associated with $\mathcal{E}$ at the exist time from D [108].

We have shown that if $X = \{X_\mu,\ \mu \in M(E)\}$ is Gaussian random field with a dual $\hat{X}$ satisfying certain conditions, then its covariance is Green's function of a symmetric Markov process associated with a Dirichlet form with the "killing measure" part of Beurling–Deny decomposition [30].

We shall now consider the generalized Gaussian random field indexed by $C_0^\infty(\mathbb{R}^n)$, that is, $\{X_\varphi,\ \varphi \in C_0^\infty(\mathbb{R}^n)\}$. In this case, we assumed in Example 7.2.1 that a dual process $\hat{X} = \{\hat{X}_\varphi, \varphi \in C_0^\infty(\mathbb{R}^n)\}$ exists in order to get conditions for Markov property. Recall that X and $\hat{X}$ are biorthogonal if

$$E\left(X_\varphi \hat{X}_\psi\right) = \int_{\mathbb{R}^n} \varphi\psi\,dt$$

and $H(X) = H(\hat{X})$ where $H(X) = \overline{\mathrm{span}}\{X_\varphi,\ \varphi \in C_0^\infty(\mathbb{R}^n)\}$ and $H(\hat{X})$ is defined similarly and the closure is in $L_2(\Omega, \mathcal{F}, P)$. As before note that the elements of the RKHS are Schwartz distributions $\mathcal{D}' = ((C_0^\infty(\mathbb{R}^n))^*$ in view of the duality

$$f_\psi(\varphi) = \int_{\mathbb{R}^n} \varphi\psi(t)\,dt \in K(C_X).$$

With $\hat{M}(D) = \overline{\mathrm{span}}^{K(C)}\{f_\psi(\cdot),\ \mathrm{supp}(\psi) \subseteq D\}$ and $M(D) = \overline{\mathrm{span}}^{K(C)}\{f \in K(C_X),\ \mathrm{supp}(f) \subseteq D\}$, as before we have $\hat{M}(D) \subseteq M(D)$.

We also assume that there exists a function $w \in C_0^\infty(\mathbb{R}^n)$ satisfying

$$E\left(\hat{X}_{w\varphi}\right)^2 \leq C_w E\left(\hat{X}_\varphi\right)^2$$

analogous to condition (7.16). Under this condition, we obtain the equality $M(D) = \hat{M}(D)$, and we conclude that $X = \{X_\varphi,\ \varphi \in C_0^\infty(\mathbb{R}^n)\}$ has GFMP if and only if $\hat{X} = \{\hat{X}_\varphi,\ \varphi \in C_0^\infty(\mathbb{R}^n)\}$ satisfies the condition that if $\varphi, \psi \in C_0^\infty(\mathbb{R}^n)$, with $\mathrm{supp}(\varphi) \cap \mathrm{supp}(\psi) = \varnothing$ then $E\hat{X}_\varphi \hat{X}_\psi = 0$, that is, in the terminology of [37], $\hat{X}$ is a Gaussian process with independent values at every point. By Theorem 9, p. 287 in [37], we obtain that its covariance is given by

$$C_{\hat{X}}(\varphi, \psi) = \int_{\mathbb{R}^n} \sum_{j,k} a_{j,k}(x) D^{(j)} \varphi(x) D^{(k)} \psi(x)\,dx,$$

where only a finite number of the coefficients $a_{j,k}(x)$ are different from zero on a given bounded set. Let us consider on $C_0^\infty(\mathbb{R}^n)$, the Dirichlet form

$$\mathcal{E}(\varphi, \psi) = C_{\hat{X}}(\varphi, \psi).$$

Using the fact that $C_0^\infty(\mathbb{R}^n)$ is dense in $L_2(\mathbb{R}^n)$ we can extend this Dirichlet form to $L^2(\mathbb{R}^n)$ and note that $\mathcal{E}$ is local. For $\varphi, \psi \in C_0^\infty(\mathbb{R}^n)$, we have

$$\int_{\mathbb{R}^n} \varphi(t)\psi(t)\,dt = E\left(\hat{X}_\varphi X_\psi\right) \leq \left(E\left(\hat{X}_\varphi^2\right)\right)^{1/2}\left(E\left(X_\psi^2\right)\right)^{1/2} \tag{8.9}$$

$$\leq \left(\mathcal{E}(\varphi, \varphi)\right)^{1/2}\left(C_X(\psi, \psi)\right)^{1/2}, \tag{8.10}$$

where C_X is the covariance of X.

Hence, the measures $\mu(A) = \int_A \varphi \, dt$ are of finite energy. We denote by U_μ the potential of μ and observe that for all $\psi \in C_0^\infty(\mathbb{R}^n)$

$$E\left(\hat{X}_{U_\mu} \hat{X}_\psi\right) = \int_E \tilde{\varphi}(t)\psi(t)\,dt = EX_\varphi \hat{X}_\psi$$

with $\tilde{\varphi}$ denoting a quasi-continuous modification (in restricted sense) of φ.

In view of the fact that $H(X) = H(\hat{X})$ we have the equality $\hat{X}_{U_\mu} = X_\varphi$ for all $\mu = \varphi \, dt$, and as a consequence, for $\mu = \varphi \, dt$ and $\nu = \psi \, dt$

$$C_X(\varphi, \psi) = E\left(\hat{X}_\mu \hat{X}_\nu\right).$$

That is, the covariance of the Gaussian random field $\left\{X_\varphi, \; \varphi \in C_0^\infty(\mathbb{R}^n)\right\}$ is given by the potential of the Markov process associated with the local Dirichlet form $\mathcal{E}$.

Theorem 8.2.2. *A Gaussian random field $\left\{X_\varphi, \; \varphi \in C_0^\infty(\mathbb{R}^n)\right\}$ having a dual field $\left\{\hat{X}_\psi, \; \psi \in C_0^\infty(\mathbb{R}^n)\right\}$ satisfying $M(D) = \hat{M}(D)$ for all open sets (respectively, all relatively compact open sets) D has GFMP on all open sets (respectively, all relatively compact open sets) D if and only if there exists a Dirichlet form $\mathcal{E}$ with the domain $\mathcal{D}(\mathcal{E}) = C_0^\infty(\mathbb{R}^n)$, such that*

$$C_X(\varphi, \psi) = \mathcal{E}\left(U_\mu, U_\nu\right),$$

where $\mu = \varphi \, dt$ and $\nu = \psi \, dt$ and $\varphi, \psi \in C_0^\infty(\mathbb{R}^n)$, with

$$\mathcal{E}(\varphi, \psi) = \sum_{j,k} \int_{\mathbb{R}^n} a_{j,k} D^j \varphi(t) D^k \psi(t)\,dt,$$

where only finite number of the coefficients $a_{j,k}(x)$ are different from zero on a given bounded set.

Bibliography

[1] M. Abramowitz and Stegun I. A. (Eds.). *Handbook of Mathematical Functions with Formulas, Graphs, and Mathematical Tables, Orthogonal Polynomials*. Dover, New York, 1972.

[2] S. Albeverio and V. Mandrekar. Remark on Gaussian Markov random fields and Dirichlet forms. *(to appear)*.

[3] N. Aronszajn. Theory of reproducing kernels. *Trans. Amer. Math. Soc.*, 68(3):337–404, 1950.

[4] A. V. Balakrishnan. Applied functional analysis. *Applications of Mathematics*, 3, 1976.

[5] A. Bensoussan. Filtrage opimal des systémes linéaires. In *Methodes*. Dunod, 1971.

[6] J.M. Bismut. Calcul des variations stochastiques et processus de sauts. *Z. für Wahrscheinlichkeitstheorie verw. Gebiete*, 63:147–235, 1983.

[7] C. Borel. Gaussian random measures on locally convex space. *Math. Scand.*, 38:265–284, 1976.

[8] R. Buckdahn. Transformations on the Wiener space and Skorohod-type stochastic differential equations. *Seminarbericht, Sektion Mathematik*, 105, 1989.

[9] R. Buckdahn. Anticipative Girsanov transformations. *Probab. Th. Rel. Fields*, 89:211–238, 1991.

[10] R.H. Cameron and W.T. Martin. Transformations of Wiener integrals under translations. *Ann. Math.*, 45:386–396, 1944.

[11] T. Carleman. Über die Abelsche Integralgleichung mit konstanten Integrationsgrenzen. *Math. Z.*, 15:111–120, 1922.

[12] S.D. Chatterji. Les martingales et leurs applications analytiques. *Ecole d'Été de Probabilités, Lecture Notes Math., (Bretagnolle L. J. et al. eds.)*, 307:27–164, 1973.

[13] S.D. Chatterji and V. Mandrekar. Quasi-invariance of measures under translations. *Math. Zeit.*, 154:19–29, 1977.

[14] S.D. Chatterji and V. Mandrekar. Singularity and absolute continuity of measures. *North-Holland Mathematics Studies*, 27:247–257, 1977.

[15] S.D. Chatterji and V. Mandrekar. Equivalence and singularity of Gaussian measures and applications. *Probabilistic Analysis and Related Topics, (Bharucha-Reid ed.)*, 1:169–197, 1978.

[16] H. Cramér. A contribution to the theory of stochastic processes. In *Proc. 2nd Berkeley Symp. on Math. Statist. and Probab.*, pages 329–339. University of California Press, 1951.

[17] J. Diestel and J.J. Uhl. Vector measures. *Mathematical Surveys*, 15, 1977.

[18] J.L. Doob. *Stochastic Processes*. Wiley, 1953.

[19] J. Du. *Asymptotic and Computational Methods in Spatial Statistics*. Ph.D. Thesis at Michigan State University, 2009.

[20] R. Durrett. *Probability: Theory and Examples*. Cambridge University Press, 2010.

[21] E.B. Dynkin. Markov processes and random fields. *Bull. Amer. Math. Soc. (N.S.)*, 3(3):975–999, 1980.

[22] E.B. Dynkin and A. Mandelbaum. Symmetric statistics, Poisson point process and multiple Wiener integrals. *Ann. Statist.*, 11:739–745, 1983.

[23] O. Enchev. Nonlinear transformations on the Wiener space. *Ann. Probab.*, 21:2169–2188, 1993.

[24] L. Evans and R. Gariepy. *Measure Theory and Fine Properties of Functions. Studies in Advanced Mathematics*. CRC Press, Boca Raton, FL, 1992.

[25] H. Federer. *Geometric Measure Theory*. Die Grundlehren der mathematischen Wissenschaften 153. Springer-Verlag, New York, 1969.

[26] J. Feldman. Some classes of equivalent Gaussian processes on an interval. *Pacific J. Math.*, 10:1211–1220, 1960.

[27] J. Feldman. Examples of non-Gaussian quasi-invariant distributions in Hilbert space. *Trans. Amer. Math. Soc.*, 99:342–349, 1961.

[28] D. Filipović. Consistency problems for Heath-Jarrow-Morton interest rate models. *Lecture Notes in Math.*, 1760, 2001.

[29] H. Föllmer. Calcul d'Itô sans Probabilités. *Séminaire de Probabilités XV, Lecture Notes in Math.*, 850:143–150, 1981.

[30] M. Fukushima, M. Takeda, and Y. Oshima. *Dirichlet Forms and Symmetric Markov Processes*. De Gruyter, 1994.

[31] L. Gawarecki and V. Mandrekar. Itô-Ramer, Skorohod and Ogawa integrals with respect to Gaussian processes and their interrelationship. In *Proceedings of the Conference on Chaos Expansions, Multiple Wiener-Itô Integrals and Their Applications, Guanajuato, Mexico, Probab. Stochastics Ser.*, pages 349–373. CRC Press, Boca Raton, FL, 1994.

[32] L. Gawarecki and V. Mandrekar. On Girsanov type theorem for antici-pative shifts. *Probability in Banach Spaces, 9. Progress in Probability* (J. Hoffman-Jørgensen et al., eds.), 35:301–316, 1994.

[33] L. Gawarecki and V. Mandrekar. On the Zakai equation of filtering with Gaussian noise. *Stochastics in Finite and Infinite Dimensions, Trends Math.*, pages 145–151, 2001.

[34] L. Gawarecki and V. Mandrekar. Remark on instrumentation problem of A.V. Balakrishnan. *Journal of the Indian Statistical Association, Special Issue in Honor of Professor S.R. Adke*, 41(2):275–284, 2003.

[35] L. Gawarecki and V. Mandrekar. Non-linear filtering with Gaussian martingale noise: Kalman filter with fBm noise. *A Festschrift for Her-man Rubin, IMS Lecture Notes Monogr.*, 45:92–97, 2004.

[36] L. Gawarecki and V. Mandrekar. *Stochastic Differential Equations in Infinite Dimensions*. Springer, Berlin, 2011.

[37] I.M. Gelfand and N.Y. Vilenkin. *Generalized Functions*, volume 4. Aca-demic Press, New York, 1964.

[38] I. Gohberg, S. Goldberg, and N. Krupnik. Traces and determinants of linear operators. *Integr. Equat. Oper. Th.*, 26, 1996.

[39] S.E. Graversen and M. Rao. Quadratic variation and energy. *Nagoya Math. J.*, 100:163–180, 1985.

[40] L. Gross. Abstract Wiener spaces. *Proc. 5th. Berkeley Sym. Math. Stat. Prob.*, 2:31–42, 1965.

[41] L. Gross. Potential theory on Hilbert space. *J. Func. Anal.*, 1:123–181, 1967.

[42] J. Hajek. On a property of normal distribution of any stochastic process. *Math. Statist. Prob.*, 1:245–252, 1958-1961.

[43] D. Heath, R. Jarrow, and A. Morton. Bond pricing and the term structure of interest rates: A new methodology for contingent claims valuation. *Econometrica*, 40(1):77–105, 1992.

[44] W. Hoeffding. A class of statistics with asymptotically normal distribu-tions. *Ann. Statist.*, 19:293–325, 1948.

[45] S. T. Huang and S. Cambanis. Stochastic and multiple Wiener integrals for Gaussian processes. *Ann. Probab.*, 6(4):585–614, 1978.

[46] I.A. Ibragimov and Y.I. Rozanov. *Gaussian Random Processes*. Springer Verlag, New York, 1978.

[47] K. Inoue. Equivalence of measures for some class of Gaussian random fields. *J. Multivariate Anal.*, 6:295–308, 1976.

[48] K. Itô. Multiple Wiener integral. *J. Math. Soc. Japan.*, 3:157–169, 1951.

[49] L. Kai. Stability of infinite dimensional stochastic differential equations

with applications. In *Monographs and Surveys in Pure and Applied Mathematics*, volume 135. Boca Raton: Chapman & Hall/CRC, 2006.

[50] T. Kailath. On measures equivalent to Wiener measure. *Ann. Math. Statist.*, 38:261–263, 1967.

[51] S. Kakutani. Spectral analysis of stationary Gaussian processes. In *Proc. 4th Berkeley Symp.*, volume 2, pages 239–247, 1961.

[52] G. Kallianpur. *Stochastic Filtering Theory*. Springer-Verlag, New York, 1980.

[53] G. Kallianpur and R.L. Karandikar. *White Noise Theory of Prediction, Filtering and Smoothing*. Gordon and Breach, New York, 1988.

[54] G. Kallianpur and R.L. Karandikar. *Nonlinear Transformations of the Canonical Gauss Measure on Hilbert Space and Absolute Continuity, Technical Report No. 387*. University of North Carolina, Chapel Hill, 1993.

[55] G. Kallianpur and H. Oodaira. The equivalence and singularity of Gaussian measures. In *Proc. Symp. on Time Series Analysis*, pages 279–291. Wiley, New York, 1963.

[56] R.L. Karandikar. On pathwise stochastic integration. *Stochastic Process. Appl.*, 57(1):11–18, 1995.

[57] P.C. Kettler, F. Proske, and M. Rubtsov. Sensitivity with respect to the yield curve: Duration in a stochastic setting. In *Inspired by Finance*, pages 363–385. Springer, 2014.

[58] M.L. Kleptsyna, A. Le Breton, and M.C. Roubaud. Parameter estimation and optimal filtering for fractional type stochastic systems. *Stat. Inference Stoch. Process.*, 3:173–182, 2000.

[59] H. Komatsu. Ultradistribution I. Structure theorem and a characterization. *J. Fac. Sci. Univ. Tokyo IA*, 20:25–105, 1973.

[60] H. Komatsu. Ultradistributions and hyperfunctions. *Lecture Notes in Math.*, 287:163–179, 1973.

[61] H. Komatsu. Ultradistribution II. The kernel theorem and ultradistribution with support in a submanifold. *J. Fac. Sci. Univ. Tokyo IA*, 24:607–628, 1977.

[62] S. Kotani. On Markov property for stationary Gaussian processes with a multidimensional parameter. *Lecture Notes in Math.*, 330:239–250, 1974.

[63] S. Kotani and Y. Okabe. On a Markovian property of stationary Gaussian processes with a multidimensional parameter. *Lecture Notes in Math.*, 287:153–163, 1973.

[64] I. Kruk, F. Russo, and C.A. Tudor. Wiener integrals, Malliavin calculus and covariance measure structure. *J. Funct. Anal.*, 249:92–142, 2007.

[65] J. Kuelbs. Gaussian measure on a Banach space. *J. Funct. Anal.*, 5:354–367, 1970.

[66] H. Künch. Gaussian Markov random fields. *J. Fac. Science University Tokyo*, Ser. A 1, Mat. 26:53–73, 1979.

[67] H. Kunita. In *A Festschrift in Honour of Gopinath Kallianpur,* S. Cambanis et al., eds., pages 201–210. Springer-Verlag, New York.

[68] H-H. Kuo. Gaussian measures in Banach spaces. *Lecture Notes in Math.*, 463, 1975.

[69] S. Kusuoka. Markov fields and local operators. *J. Fac. Science University Tokyo*, Ser. A 1:199–212, 1979.

[70] S. Kusuoka. The non-linear transformation of Gaussian measure on Banach space and its absolute continuity (I). *J. Fac. Science University Tokyo*, Ser. A 1:575–597, 1982.

[71] A. Le Breton. A Girsanov-type approach to filtering in a simple linear system with fractional Brownian noise. *C. R. Acad. Sci. Paris Ser. I Math.*, 326:997–1002, 1998.

[72] W.E. Leland, M.S. Taqqu, W. Willinger, and D.V. Wilson. On the self-similar nature of ethernet traffic (extended version). *IEEE/ACM Trans. Networking*, 2:1–15, 1994.

[73] P. Lévy. A special problem of Brownian motion, and a general theory of Gaussian random functions. In *Proceedings of the Third Berkeley Symposium on Mathematical Statistics and Probability, Contributions to Probability Theory*, volume 2, pages 133–175. University of California Press, 1956.

[74] R.S. Liptser and A.N. Shiryaev. *Statistics of Random Processes.* Springer-Verlag, New York, 1977.

[75] F.R. Macaulay. *Some Theoretical Problems Suggested by the Movements if Interest Rates, Bond Yields, and Stock Prices in the United States since 1856.* Columbia University Press, New York, 1938.

[76] P. Malliavin. Stochastic calculus of variations. In *Proceedings, International Conference on SDE, Kyoto*, pages 243–310, 1976.

[77] P. K. Mandal and V. Mandrekar. A Bayes formula for Gaussian noise processes and its applications. *SIAM J. Control Optim.*, 39(3):852–871, 2000.

[78] A. Mandelbaum and M.S. Taqqu. Invariance principles for symmetric statistics. *Ann. Statist.*, 12:483–496, 1984.

[79] V. Mandrekar. Gaussian processes and Markov property. *Lecture Notes at Ecole Polytechnique Lausanne (unpublished).*

[80] V. Mandrekar. Stochastic integral with respect to Gaussian processes. *Lecture Notes in Math.*, 1089:288–293, 1984.

[81] V. Mandrekar and G. Kallianpur. The Markov property for generalized Gaussian random fields. *Ann. Inst. Fourier (Grenoble)*, 24:143–167, 1974.

[82] V. Mandrekar and B.V. Rao. On a limit theorem and invariance principle for symmetric statistics. *Probab. Math. Statist.*, 10:271–276, 1989.

[83] V. Mandrekar and A.R. Soltani. *Markov Property for Gaussian Ultraprocesses, Technical Report 5.* Center for Stochastic Processes, Department of Statistics, University of North Carolina, Chapel Hill, 1982.

[84] V. Mandrekar and S. Zhang. Skorokhod integral and differentiation for Gaussian processes. *R.R. Bahadur Festschrift, Stat. and Prob.* (J. K. Ghosh, et al. eds.), pages 395–410, 1994.

[85] G. Matheron. Traité de Géostatistique Appliquée. *Paris: Editions Technip.*, 1, 1962.

[86] S. Mazur and W. Orlicz. On a Limit theorem and invariance principle for symmetric statistics. *Studia Math.*, 17:97–119, 1958.

[87] H. P. McKean, Jr. Brownian motion with several dimensional time. *Theor. Probab. Appl.*, 8:335–354, 1963.

[88] G.M. Molchan. Characterization of Gaussian fields with Markovian property. *Dokl. Acad. Nauk SSSR*, 197:563–567, 1971.

[89] D. Mondal. Applying Dynkin's isomorphism: an alternative approach to understand the Markov Property of the de Wijs process. *To appear in Bernoulli.*

[90] E. Nelson. The free Markov field. *J. Funct. Anal.*, 12:211–227, 1973.

[91] J. Neveu. *Mathematical Foundation of the Calculus of Probability.* Holden-Day, Inc., San Francisco, 1972.

[92] J. Neveu. *Discrete Parameter Martingales.* North Holland, Amsterdam, 1974.

[93] I. Norros, E. Valkeila, and J. Virtamo. An elementary approach to Girsanov formula and other analytical results on fractional Brownian motion. *Bernoulli*, 5:571–587, 1999.

[94] D. Nualart. Noncausal stochastic integrals and calculus. *Lecture Notes in Math.*, 1316, 1988.

[95] D. Nualart. *Malliavin Calculus and Related Topics.* Springer, New York, 2006.

[96] D. Nualart and E. Pardoux. Stochastic calculus with anticipating integrands. *Probab. Theory Related Fields*, 78:535–581, 1988.

[97] D. Nualart and M. Zakai. Generalized stochastic integrals and the Malliavin calculus. *Probab. Theory Related Fields*, 73:255–280, 1986.

[98] S. Ogawa. Sur le Produit Direct du Bruit Blanc par Lui-Méme. *C.R.*

Acad. Sci. Paris, Sér. A, 288:359–362, 1979.

[99] S. Ogawa. Quelques Propriétés de L'intégrale Stochastique du Type Noncausal. *Japan J. Appl. Math.*, 1:405–416, 1984.

[100] S. Ogawa. The stochastic integral of noncausal type as an extension of the symmetric integrals. *Japan J. Appl. Math.*, 2:229–240, 1985.

[101] Y. Okabe. Stationary Gaussian processes with Markovian property and M. Sato's hyperfunctions. *Japan J. Appl. Math.*, 41:69–122, 1973.

[102] W.J. Park. A multiparameter Gaussian process. *Ann. Math. Stat.*, 41:1582–1595, 1970.

[103] E. Parzen. Probability density functionals and reproducing kernel Hilbert spaces. In *Proc. Symp. on Time Series Analysis*, pages 155–169. Wiley, New York, 1963.

[104] L.D. Pitt. A Markov property for Gaussian processes with multi-dimensional parameter. *Arch. Rat. Mech. Anal.*, 43:367–391, 1971.

[105] L.D. Pitt. Some problems in the spectral theory of stationary processes on $\mathbb{R}^d$. *Indiana Univ. Math. J.*, 23:243–365, 1973.

[106] S.C. Port and C.J. Stone. *Brownian Motion and Classical Potential Theory*. Academic Press, New York, 1978.

[107] Ramer R. On non-linear transformations of Gaussian measures. *J. Funct. Anal.*, 15:166–187, 1974.

[108] M. Röckner. Generalized Markov fields and Dirichlet forms. *Acta Appl. Math.*, 3:285–311, 1985.

[109] J. Rosiński. On stochastic integration by series of Wiener integrals. *Appl. Math. Optim.*, 19:137–155, 1989.

[110] H.L. Royden. *Real Analysis*. MacMillan, New York, 1963.

[111] Yu.A. Rozanov. *Markov Random Fields*. Springer, New York, 1982.

[112] W. Rudin. *Real and Complex Analysis*. McGraw-Hill, New York, 1987.

[113] J.T. Schwartz. *Non-Linear Functional Analysis*. Gordon and Breach Science Publishers, 1987.

[114] R.J. Serfling. *Approximation Theorems of Mathematical Statistics*. Wiley, New York, 1980.

[115] L.A. Shepp. Radon-Nykodym derivatives of Gaussian measures. *Ann. Math. Stat.*, 37:321–354, 1966.

[116] B. Simon. *Trace Ideals and Their Applications*. Cambridge University Press, 1979.

[117] A.V. Skorokhod. On differentiability of measures which correspond to stochastic processes I. *Theory Probab. Appl.*, 2:417–443, 1957.

[118] A.V. Skorokhod. On a generalization of stochastic integral. *Theory*

Probab. Appl., 20:219–233, 1975.

[119] D. Sondermann. Introduction to stochastic calculus for finance. A new didactic approach. *Lecture Notes in Economics and Mathematical Systems*, 579, 2006.

[120] M. Stein. *Statistical Interpolation of Spatial Data: Some Theory for Kriging*. Springer, New York, 1999.

[121] J. von Neumann. On rings of operators. Reduction theory. *Ann. of Math. (2)*, 50:401–485, 1949.

[122] N. Wiener. *Non–Linear Problems in Random Theory*. MIT University Press, Cambridge, MA, 1958.

[123] M. Zakai and O. Zeitouni. When does the Ramer formula look like the Girsanov formula? *Ann. Probab.*, 20(3):1436–1449, 1992.

[124] S. Zhang. *Markov Properties of Measure–Indexed Gaussian Random Fields*. Ph.D. Thesis, Department of Statistics and Probability, Michigan State University, 1990.

Index

For Product Safety Concerns and Information please contact our
EU representative GPSR@taylorandfrancis.com Taylor & Francis
Verlag GmbH, Kaufingerstraße 24, 80331 München, Germany